SpringerBriefs in Computer Science

For further volumes:
http://www.springer.com/series/10028

Márcio Moretto Ribeiro

Belief Revision in
Non-Classical Logics

 Springer

Márcio Moretto Ribeiro
Centro de Lógica Epistemologia e História da Ciência
UNICAMP
Campinas, São Paulo
Brazil

ISSN 2191-5768   ISSN 2191-5776   (electronic)
ISBN 978-1-4471-4185-3   ISBN 978-1-4471-4186-0   (eBook)
DOI 10.1007/978-1-4471-4186-0
Springer London Heidelberg New York Dordrecht

Library of Congress Control Number: 2012941631

Printed on acid-free paper

Springer is part of Springer Science+Business Media (www.springer.com)

To Professor Renata Wassermann

Preface

This work started when I first met my adviser, Renata Wassermann, during my under-graduation and it finishes now 7 years later. In the first years of work I had the privilege of meeting professor Grigoris Antoniou who presented me the work of his student Giorgos Flouris. Me and Renata were already working with belief revision in Description Logics and Flouris had just presented their results on AGM compliance. Although extremely interesting, their results were negative. They showed essentially how AGM theory was not applicable to several interesting Description Logics such as the ones behind OWL.

We decided then to investigate how to adapt AGM theory to make it applicable to these and other logics. First, we investigated contraction in belief sets and then revision both in belief bases and belief sets. In 2010, I finished my Ph.D. thesis which was elected the best thesis in Computer Science of that year by the Brasilian Society of Computer Science. This is why it was published.

To make it look like a book the thesis suffered some modifications. Some classical results were proved to make it more self-contained. Some results were removed because they were less related with the subject.

I would like to acknowledge many people, but I will mention just some. First and most important I would like to thank Renata. In many ways this can be considered our joint work.

For funding this work I want to thank CNPq (Conselho Nacional de Desenvolvimento Científico e Tecnológico) and FAPESP (Fundação de Amparo a Pesquisa do Estado de São Paulo)

I would like to thank people from LIAMF, especially professor Marcelo Finger, professor Leliane Nunes, and professor Fabio Cozman who is not officially from LIAMF, but is always present. Some of the works that I started in LIAMF are being continued by some students who are doing a great job. I would like to mention Filipe Resina who is implementing the algorithms presented in this book and Renato Lundberg who is investigating the relation between kernel and remainder set.

I am in debt with everyone who hosted me in Europe, especially professor Griogoris Antoniou, Giorgos Flouris, and professor Andreas Herzig. I would also like to thank the people from my current work in CLE (Centro de Lógica Epistemologia e Historia da Ciência), especially my current adviser professor Marcelo Coniglio.

Finally, I would like to mention everyone who supported me all these years, my friends in Vinhedo, my friends in São Paulo, my colleagues from my undergraduation course, my family, and my beloved Bianca.

Contents

Chapter 1
Introduction

Abstract Belief revision is a mature research field that deals with the dynamics of epistemic states. It has applications in many different areas such as computer science, philosophy, and law. In the past decade, specially motivated by semantic Web applications, the interest in applying belief revision theory to non-classical logics has grown rapidly. The first results on belief revision in Description Logics (DL) showed problems in applying "classical belief revision theory" (AGM theory) to these logics. The aim of this book is to show how to adapt belief revision theory to make it applicable to a wider class of logics that includes most DLs.

Keywords Belief revision · Belief bases · Non-classical logics · Semantic web · Ontology evolution

In our everyday life we are constantly changing our mind. Some months ago, for example, I did not believe that I would write a book in a near future, but now I do. In this simple case of belief change I only needed to add a new piece of information to my beliefs. Sometimes, however, changing our beliefs can be more complicated.

Consider the following example:

Example 1.1 My friend Aristotle believed that every man is mortal. He also believed that Socrates is a man. Hence, he believed that Socrates is mortal. However, talking with her friend Hannah she convinced him that Socrates' accomplishments made him immortal.

How should Aristotle incorporate this new belief? This new piece of information certainly contradicts what he believed earlier. Hence, Aristotle has to abandon certain beliefs to accommodate the new belief that Socrates is immortal. However, it is not obvious which beliefs should he drop. He may consider Socrates a kind of demigod and abandon his belief that he is a man or that not every man is mortal for example. In the latter case he may consider that every man except certain ones are mortal. Any of these possibilities seem plausible a priori.

Furthermore, information is expensive. It would be irrational for Aristotle to abandon more beliefs than he has to. Aristotle would seem irrational if, in order to

M. M. Ribeiro, *Belief Revision in Non-Classical Logics*,
SpringerBriefs in Computer Science,
DOI: 10.1007/978-1-4471-4186-0_1, © The Author(s) 2013

accommodate this new piece of information, he abandoned his beliefs about democracy, for example.

1.1 Belief Revision

Belief revision is the research field that studies the rationality of belief change, i.e., how do rational agents change their beliefs and how should they change it. Before presenting the dynamics of belief, we need to clarify what we understand by *belief*.

When we talk about belief we do not mean belief as in "I believe in what my mother say" or in "I believe in God". We are interested in the meaning of belief as in "Aristotle believes that Socrates is mortal", i.e., belief as a relation between an agent and a proposition.

An *agent* is simply defined, following Russel and Norvig [RN03], as any entity capable of perceiving and acting on the world. A *proposition*, on the other hand, is not that easy to define. Some philosophers define it as the content of a sentence. For our purposes, a proposition is just a sentence in certain logic, i.e., a sentence in a language with a consequence operator.

Of course not every relation between an agent and a proposition is a belief. Hardly nobody would say that I believe in the sentence "I am hairy" if it was written in my forehead, although I would certainly have a relation with this proposition. Restricting which relation between an agent and a proposition properly defines belief is hard and touches several deep philosophical questions such as: Is there an explicit representation of the believe somewhere in the agent, in his mind for example? Is there a limit on how many propositions can an agent believe in? Can an agent believe contradictory propositions? What is the difference between belief and knowledge?

As a matter of fact, precisely defining belief is hard and we will let it to philosophers. In this book we just assume that at each moment the beliefs of an agent, its *epistemic state*, can be represented by certain set of sentences. Whether this set represents the cognitive state of the agent is not important for us. Furthermore, we would not make any distinction between knowledge and belief.

Now consider an agent with certain epistemic state when it hears something, or reads something, or smells something, etc., which makes it change its epistemic state. Whatever makes it change its mind is called the *input* of the change. An *operation* is the function from the old epistemic state to the new epistemic state given the input.

There are three main operations over epistemic states: *revision*, *expansion*, and *contraction*. In the beginning of this chapter we presented examples of expansion (me being informed that I would write a book) and revision (Aristotle learning that Socrates is immortal). Expansion is the operation used when some proposition is added to the agent's epistemic state. Revision is used when this addition is done in a consistent way. Contraction is used when the goal of the agent is to "open its mind" about certain proposition, i.e., when it wants not to have an opinion about the sentence.

1.2 Belief Sets and Belief Bases

There are several ways of representing the epistemic state of an agent. One can represent the beliefs of an agent as sentences with probabilities or simply as sentences in a certain logic, for example. We will consider two ways of representing an epistemic state: as a belief set and as a belief base. In both ways an epistemic state is represented as a set of sentences and both assume that the agent respects certain logic, i.e., there is a notion of consequence underlying the agent's beliefs. The first represents the epistemic state of an agent as a set of sentences closed under the consequence operation called belief set. The second represents the epistemic state as an arbitrary set of sentences called belief base. Each of these representations have advantages as well as disadvantages.

In belief set representation the syntactic form of the epistemic state is abstracted away, i.e., the form of the sentences do not matter, all that matters is their semantic content. Certain philosophers defend this representation arguing that there is no way to distinguish the syntactic form of an agent's belief.

Belief base representation, on the other hand, distinguishes explicit beliefs from implicit beliefs which can be abandoned more easily without cluttering the agent's mind. Certain philosophers defend this representation arguing that it is more expressive and more economic. Furthermore, this representation clearly has advantages in computational complexity.

1.3 Postulates and Constructions

The most common way of defining an operations over an epistemic state is via a set of *rationality postulates*. Rationality postulates state criteria that should be satisfied by a rational agent when it performs the operation. Let us assume that we are representing the agent's epistemic state as a belief set. *Success* of contraction over a belief set states that, after performing the contraction, the resulting belief set should not contain the input, *closure* states that it should be logically closed, etc.

The most influential work in belief revision is [AGM85]. In this work, the authors defined rationality postulates for revision and contraction over belief sets. These sets of postulates are called AGM postulates due to their authors initials.

Besides presenting these postulates, the authors also presented the operations constructively. The construction for contraction presented in [AGM85], partial meet contraction, is fully characterized by the AGM postulates, i.e., every partial meet contraction satisfies the AGM postulates for contraction and every contraction that satisfies the AGM postulates is a partial meet contraction. After AGM's work, many constructions for revision and contraction were presented, e.g.: Grove system of spheres [Gro88], safe contraction [Rot92], and epistemic entrenchment [Gär88]. All of them were proved equivalent to the AGM postulates.

The result showing that certain construction is fully characterized by a set of postulates is central in AGM theory and is called *representation theorem*. Once a representation theorem is proved one can examine the construction by studying the postulates that characterizes it. This way the details of implementation can be abstracted away. If one wants to criticize the rationality of the operation she should criticize some of the postulates rather than the construction itself. Chapter 4 presents classical belief revision theory. Both belief base and belief set representations are presented together with constructions and rationality postulates for the main operations.

1.4 Non-Classical Logics

We already argued that an agent respects some logic, but why don't every agent respect one and the same logic? Why don't they all respect classical propositional logic (CPL) or first-order logic, for example?

Following [Eps90], non-classical logics are used whenever one wants to emphasize certain aspects of reasoning. For example, it is useful to use temporal logics when time is the aspect to be emphasized. Furthermore, as pointed by Levesque and Brachman [LB87], in computational applications it is useful to use less expressive logics in order to keep reasoning computationally more efficient (less complex). Summing up, one should choose the most adequate formalism to represent an agent's belief.

For example, Description Logics (DLs) have been pointed as an adequate formalism to represent conceptual knowledge (ontologies). Compared with their predecessors (systems of frames [Min81] and semantic networks [Qui67]), DLs have the advantage of having formal semantics based in first-order logic. Furthermore, compared with first-order logic itself, DLs have the advantage of being decidable.

In the past 20 years many DLs with different expressivity and complexity have been studied. These studies converged with the need for an ontology language for the Web. The view of a Web where artificial agents who could exchange their knowledge online needed a standard language to represent the background taxonomy of the agents terms, i.e., to represent *ontologies* [BLHL01]. The outcome was the adoption in 2004 of OWL, a language based on DLs, as the W3C standard to represent ontologies in the Web.

Another example of non-classical logic is Horn logic. Horn logic is a fragment of propositional logic. Although less expressive, the main inference problem in Horn logic is polynomial in time while the main inference problem in CPL is CoNP-complete. The efficiency of Horn logic made it an important choice in logical programming. Some important programming languages, such as PROLOG, are based in Horn logic.

Details of each of these logics can be abstracted away considering just the properties that their consequence operator satisfies. In Chap. 2 consequence operator and certain properties that it may satisfy are presented. In Chap. 3 some examples of logics, such as Horn logic and DLs, are also presented.

1.5 Belief Revision in Non-Classical Logics

Since the Web is a dynamic environment, it would not be wise to assume that ontologies in the Web would be static. For this reason, in the past few years some works focused on the study of ontology dynamics [Sto04, HS04].

However, there are several obstacles in applying AGM theory to DLs. As pointed by Flouris [Flo06], AGM postulates for contraction are not compliant with many DL including the ones behind the many flavors of OWL.

Furthermore, many applications in Aritificial Inteligence are written in PROLOG. Some of these applications uses PROLOG to represent agent's knowledge base which may be dynamic. However, AGM theory is also not applicable to Horn logics [LSST08]. In fact, AGM postulates are not compliant with several non-classical logics.

In this book we intend to show how to use belief revision techniques in logics where AGM theory is not applicable. Our main goal is to show that, although AGM theory fails in those logics, *belief revision is still a good theoretical framework to deal with belief dynamics in non-classical logics.*

Chapters 5, 6, and 7 are reserved for belief revision in non-classical logics. The first and the second present revision and contraction in belief sets, respectively. The third presents belief revision in belief bases.

1.6 Organization of the Book

We will assume from the reader certain familiarity with logic. Besides that, Chap. 2 introduces basic aspects of Tarskian logics and consequence operators. The chapter begins defining basic concepts and ends defining logical properties that will be needed in further chapters.

Chapter 3 presents a list of logics that will be used as examples for the application of the theory presented afterwards. For each of these logics it is showed the properties that its consequence operator satisfies. CPL is presented as a canonical example of well-behaved logic. Intuitionistic logic, Horn logic, and DLs are presented as examples of logic where classical belief revision theory does not apply.

Chapter 4 introduces classical belief revision theory. It presents belief revision theory both in belief sets and in belief bases. Constructions for the most important operations are defined as well as postulates that characterizes them. Representation theorems proving these characterizations are presented. These theorems hold in logics that satisfies certain assumptions which include CPL, but do not include the rest of the logics listed in Chap. 3.

Chapter 5 is dedicated to belief set contraction in non-classical logics. Two postulates for belief contraction, namely, recovery and relevance, are compared. It is argued that relevance is better suited for contraction in non-classical logics.

Chapter 6 deals with belief set revision. It presents a construction for revision in belief sets which does not depend on the underlying logic being closed under classical negation.

In Chap. 7 several constructions for belief base revision that do not depend on negation are introduced. For each construction a set of postulates that characterizes it is presented and the respective representation theorem is proved.

Chapter 8 shows algorithms for belief base contraction. Most of these algorithms were already known in other areas such as diagnosis and ontology debugging. They are presented and analyzed in the context of belief base contraction.

Chapter 9 concludes the book summing up the main results.

References

[AGM85]   Carlos Alchourrón, Peter Gärdenfors, and David Makinson. On the logic of theory change. *Journal of Symbolic Logic*, 50(2):510–530, 1985.

[BLHL01] Tim Berners-Lee, James Hendler, and Ora Lassila. The Semantic Web: A new form of web content that is meaningful to computers will unleash a revolution of new possibilities. *Scientific American*, Maio, 2001.

[Eps90]   R. L. Epstein. *The Semantics Foundations of Logic. Volume 1: Propositional Logics.* Kluwer, Dordrecht, Netherlands, 1990.

[Flo06]   Giorgos Flouris. *On Belief Change and Ontology Evolution.* PhD thesis, University of Crete, 2006.

[Gär88]   Peter Gärdenfors. *Knowledge in Flux: Modeling the Dynamics of Epistemic States.* MIT Press, Cambridge, UK, 1988.

[Gro88]   Adam Grove. Two modellings for theory change. *Journal of Philosophical Logic*, 17:157–170, 1988.

[HS04]   Peter Haase and York Sure. State-of-the-art on ontology evolution. SEKT informal deliverable 3.1.1.b, Institute AIFB, University of Karlsruhe, 2004.

[LB87]   Hector J. Levesque and Ronald J. Brachman. Expressiveness and tractability in knowledge representation and reasoning. *Computational Intelligence*, 3:78–93, 1987.

[LSST08] Marina Langlois, Robert H. Sloan, Balázs Szörényi, and György Turán. Horn complements: Towards Horn-to-Horn belief revision. In Dieter Fox and Carla P. Gomes, editors, *Proceedings of the 23rd AAAI Conference on Artificial Intelligence (AAAI 2008)*, pages 466–471, Chicago, Illinois, USA, Julho 13-17 2008. AAAI Press.

[Min81]   Marvin Minsky. A framework for representing knowledge. In John Haugeland, editor, *Mind Design: Philosophy, Psychology, Artificial Intelligence*, pages 95–128. MIT Press, Cambridge, MA, 1981.

[Qui67]   Ross M. Quillian. Word concepts: A theory and simulation of some basic semantic capabilities. *Behavioral Science*, 12:410–430, 1967.

[RN03]   Stuart Russell and Peter Norvig. *Artificial Intelligence: A Modern Approach.* Prentice-Hall, Englewood Cliffs, NJ, 2 edition, 2003.

[Rot92]   Hans Rott. On the logic of theory change: More maps between different kinds of contraction functions. In Peter Gärdenfors, editor, *Belief Revision*. Cambridge University Press, 1992.

[Sto04]   Ljiljana Stojanovic. *Methods and Tools for Ontology Evolution.* PhD thesis, University of Karlsruhe, Germany, 2004.

Chapter 2
Consequence

Abstract Belief revision theory assumes that the agents reason according to a logic. As an abstraction that encompasses many different logics, we can consider a logic as a pair $\langle \mathscr{L}, Cn \rangle$ such that $\mathscr{L}$ is the *language* of the logic and Cn is the *consequence operation* $Cn : 2^{\mathscr{L}} \to 2^{\mathscr{L}}$ that gives the consequences of a set of sentences. We are particularly interested in Tarskian logics and certain properties that they may satisfy e.g., compactness, decomposability, distribuitivity, etc. In this chapter, Tarskian logics, some logical properties, and relations between these properties will be presented.

Keywords Consequence operator · Consequence relation · Tarskian logics · Lattices

As usual, the symbol $2^{\mathscr{L}}$ represent the set of subsets of $\mathscr{L}$. We represent sets of sentences (subsets of $\mathscr{L}$) using upper case letters $A, B, C, \ldots$. Sentences of the language are represented with lower case Greek letters $\alpha, \beta, \ldots$.

Given a logic $\langle \mathscr{L}, Cn \rangle$, consider a sentence $\alpha \in \mathscr{L}$ and two sets of sentences $A, B \in 2^{\mathscr{L}}$:

1. α is a *consequence* of B iff $\alpha \in Cn(A)$.
2. A is a *consequence* of B iff every element of A is a consequence of B i.e., $A \subseteq Cn(B)$.
3. A and B are *equivalent* iff $Cn(A) = Cn(B)$.
4. A is *trivial* iff $Cn(A) = \mathscr{L}$.

Following the above definitions, a set A is *not* a consequence of B iff $A \not\subseteq Cn(B)$ i.e., there is at least one sentence $\alpha \in A$ which is not a consequence of B. An alternative definition would impose that A is not a consequence of B iff $B \cap Cn(A) = \emptyset$. We will use the former.

M. M. Ribeiro, *Belief Revision in Non-Classical Logics*,
SpringerBriefs in Computer Science,
DOI: 10.1007/978-1-4471-4186-0_2, © The Author(s) 2013

2.1 Tarskian Consequence Operator

Some very basic properties that a consequence operator may satisfy are the following:

monotonicity: if $A \subseteq B$ then $Cn(A) \subseteq Cn(B)$.
idempotence: $Cn(A) = Cn(Cn(A))$.
inclusion: $A \subseteq Cn(A)$.

A consequence operator that satisfies these properties is called *Tarskian*. In general, if a consequence operator of a logic satisfies certain property, we will simply say that the logic itself satisfies it. For example, a logic $\langle \mathscr{L}, Cn \rangle$ with a Tarskian consequence operator is also called *Tarskian*.

Although Tarskian logics do not encompass every logic in the literature (e.g., linear logic [Gir87] and non-monotonic logics [AW97] are not Tarskian), it encompasses enough logics so that we will only consider them in this book. Hence, from now on, whenever we write logic we mean Tarskian logic.

The following are simple lemmas about Tarskian consequence operator that will be used throughout the book without reference:

Lemma 2.1 *Let A and B be sets of sentences in a Tarskian logic $\langle \mathscr{L}, Cn \rangle$. Then the following equations hold*:

1. $Cn(Cn(A) \cup Cn(B)) = Cn(A \cup B)$
2. $Cn(Cn(A) \cap Cn(B)) = Cn(A) \cap Cn(B)$

Proof

1. By *inclusion* we have $A \subseteq Cn(A)$ and $B \subseteq Cn(B)$. It follows that $A \cup B \subseteq Cn(A) \cup Cn(B)$ and by *monotonicity* $Cn(A \cup B) \subseteq Cn(Cn(A) \cup Cn(B))$.
   By *monotonicity* both $Cn(A)$ and $Cn(B)$ are subsets of $Cn(A \cup B)$. It follows that $Cn(A) \cup Cn(B) \subseteq Cn(A \cup B)$ and by *idempotence* $Cn(Cn(A) \cup Cn(B)) \subseteq Cn(A \cup B)$.
2. Of course, $Cn(A) \cap Cn(B) \subseteq Cn(A)$ and by *idempotence* $Cn(Cn(A) \cap Cn(B)) \subseteq Cn(A)$. Analogously we have that $Cn(Cn(A) \cap Cn(B)) \subseteq Cn(B)$. It follows that $Cn(Cn(A) \cap Cn(B)) \subseteq Cn(B) \cap Cn(A)$ $Cn(A) \cap Cn(B) \subseteq Cn(Cn(A) \cap Cn(B))$ follows directly by *inclusion*.   □

Lemma 2.2 *Let $A, B, K \in 2^{\mathscr{L}}$. If A and B are equivalent then $A \subseteq Cn(K)$ iff $B \subseteq Cn(K)$.*

Proof By *monotonicity* and *idempotence* $Cn(A) \subseteq Cn(K)$. Since by hypothesis $Cn(A) = Cn(B)$ then $Cn(B) \subseteq Cn(K)$ and by *inclusion* $B \subseteq Cn(K)$.
The converse is analogous.   □

A set $K \subseteq \mathscr{L}$ is *closed* (*under* Cn) iff $K = Cn(K)$. We reserve the uppercase letter K to represent closed sets.

The class of all closed sets in a logic $\langle \mathscr{L}, Cn \rangle$ will be denoted $\mathbb{K}_{\mathscr{L}}$ or simply $\mathbb{K}$ when the context is clear. In symbols:

Fig. 2.1 This diagram represents the logic of example 2.3

$$\mathbb{K} = \{K \in 2^{\mathscr{L}} : K = Cn(K)\}$$

Of course, the relation of inclusion is a partial order over $\mathbb{K}$ i.e., for every $K_1, K_2, K_3 \in \mathbb{K}$ we have:

transitivity: if $K_1 \subseteq K_2$ and $K_2 \subseteq K_3$ then $K_1 \subseteq K_3$.
reflexivity: $K_1 \subseteq K_1$.
anti-symmetry: if $K_1 \subseteq K_2$ and $K_2 \subseteq K_1$ then $K_1 = K_2$.

For every $\Gamma \subseteq \mathbb{K}$, an *upper-bound* of Γ is any $K \in \mathbb{K}$ such that $K' \subseteq K$ for every $K' \in \Gamma$. The *supremum* or the *least upper-bound* of Γ (denoted $sup(\Gamma)$) is an upper-bound of Γ such that $sup(\Gamma) \subseteq K'$ for every upper-bound $K' \in \mathbb{K}$ of Γ. Notice that for every $\Gamma \in \mathbb{K}$, there is a unique $sup(\Gamma) \in \mathbb{K}$ defined by the formula:

$$sup(\Gamma) = Cn(\bigcup \Gamma)$$

The *infimum*, the greatest lower bound, of Γ in $\mathbb{K}$ is defined analogously. Again notice that $inf(\Gamma) = Cn(\bigcap \Gamma) \in \mathbb{K}$. Hence, $\langle \mathbb{K}, \subseteq \rangle$ form a *complete lattice* i.e., a partially ordered set such that every subsets has a supremum and an infimum.

Furthermore, $\mathbb{K}$ is closed under intersection i.e., for every $\Gamma \subseteq \mathbb{K}$ we have $\bigcap \Gamma \in \mathbb{K}$. In other words, $\mathbb{K}$ forms a *closure system*. Hence, we have that:

$$inf(\Gamma) = \bigcap \Gamma$$

Logics will be illustrated by means of diagrams like the one in Example 2.1. Each node in the diagram represent an element K of $\mathbb{K}$ and will be labeled by one set A such $Cn(A) = K$. The transitive closure of the arrows represent the consequence relation i.e., $Cn(K_j) \subseteq Cn(K_i)$ iff there is a path in the diagram from K_i to K_j.

Example 2.3 defines a simple logic[1] which is represented in the diagram of Fig. 2.1.

Example 2.3

$$\mathscr{L} = \{a, b\}$$
$$Cn(\mathscr{L}) = Cn(\{b\}) = \mathscr{L}$$
$$Cn(\{a\}) = \{a\}$$
$$Cn(\emptyset) = \emptyset$$

---

[1] This logic was borrowed from [Flo06].

Since we will present several examples of logics like the above, some conventions are useful. We will use $Cn(a)$ as an abbreviation for $Cn(\{a\})$. Furthermore, we will use a instead of $\{a\}$ to label the diagram. The value of $Cn(\mathscr{L})$ will be omitted in future presentation of logics, since it is always equal to $\mathscr{L}$ in Tarskian logics.

2.2 Consequence Relation

Another way to present Tarskian logics is via a consequence relation $R \subseteq 2^{\mathscr{L}} \times \mathscr{L}$. In this case a logic is a pair $\langle \mathscr{L}, R \rangle$ and $(A, \alpha) \in R$ means that α is a consequence of A. Tarskian consequence relation satisfies the following properties:

1. if $\alpha \in A$ then $(A, \alpha) \in R$,
2. if $(A, \alpha) \in R$ and $(B, \beta) \in R$ for every $\beta \in A$ then $(B, \alpha) \in R$ and
3. if $(A, \alpha) \in R$ and $A \subseteq B$ then $(B, \alpha) \in R$.

A Tarskian relation R induces the following consequence operator:

$$Cn(A) = \{\alpha \in \mathscr{L} : (A, \alpha) \in R\}$$

We call Cn the consequence operator that is induced by R. As we should expect, the consequence operator Cn induced by a Tarskian relation R is Tarskian:

Proposition 2.4 *If R is a Tarskian relation then the consequence operator Cn induced by R is Tarskian.*

Proof The proof is simple and follows like this:

Inclusion: If $\alpha \in A$ then $(A, \alpha) \in R$ and by definition $\alpha \in Cn(A)$.
Idempotence: Let $\alpha \in Cn(Cn(A))$. For every $\beta \in Cn(A)$, by definition, we have that $(A, \beta) \in R$. Since $(Cn(A), \alpha) \in R$ then $(A, \alpha) \in R$. Hence, $\alpha \in Cn(A)$.
Monotonicity: Let $A \subseteq B$ and $\alpha \in Cn(A)$. By definition $(A, \alpha) \in R$ which implies $(B, \alpha) \in R$. It follows that $\alpha \in Cn(B)$. $\square$

Moreover, a Tarskian consequence operator Cn induces a relation:

$$(A, \alpha) \in R \text{ iff } \alpha \in Cn(A)$$

Proposition 2.5 *If Cn is a Tarskian operator then the consequence relation R induced by Cn is Tarskian.*

Proof This proof is omitted since it is trivial and very similar to the previous one. $\square$

2.2.1 Properties of the Consequence Operator

In this section, we present a list of properties which are not as basic as the ones presented in previous section. These properties would not be generally assumed, but they will be useful throughout the book. The list of properties that will be considered include *compactness, finiteness, Descending Chain Condition (DCC), closure under complement, distribuitivity* and *decomposability*.

Compactness guaranties that any consequence of a set of sentences is a consequence of a finite subset of it:

compactness: A logic $\langle \mathcal{L}, Cn \rangle$ is *compact* iff for all $\alpha \in \mathcal{L}$ and all $A \subseteq \mathcal{L}$, if $\alpha \in Cn(A)$ then there is a finite $A' \subseteq A$ such that $\alpha \in Cn(A')$.

A consequence relation is *compact* if for every $\alpha \in \mathcal{L}$ and $A \subseteq \mathcal{L}$ we have that $(A, \alpha) \in R$ there is a finite $A' \subseteq A$ such that $(A', \alpha) \in R$. It is trivial to verify that a compact relation induces a compact consequence operator and vice-versa.

Compact logics are sometimes called finitary. We would not name them finitary to avoid confusion with what we call finite logics:

finiteness: A logic $\langle \mathcal{L}, Cn \rangle$ is *finite* iff there is only a finite number of distinct belief sets i.e., $\mathbb{K}_{\mathcal{L}}$ is finite.

Of course, finite logics are compact, but the converse is not true in general.

Another logical properties related to compactness and finiteness are the chain conditions. A sequence of sets of sentences $A_0, A_1 \dots$ is an *descending chain* iff $Cn(A_0) \supset Cn(A_1) \supset \dots$. The Descending Chain Conditions (DCC) states that any chain in $\langle \mathcal{L}, Cn \rangle$ is finite.

descending chain condition: A logic $\langle \mathcal{L}, Cn \rangle$ satisfies the *descending chain condition* iff every descending chain in $\langle \mathcal{L}, Cn \rangle$ has a minimal element i.e., a A_j such that for every A_i, we have that $Cn(A_j) \subseteq Cn(A_i)$.

It is trivial to notice that finite logics satisfies DCC.

The *complement* of a set $A \subseteq \mathcal{L}$, if it exists, is a set $A' \subseteq \mathcal{L}$ such that:

- $Cn(A \cup A') = \mathcal{L}$
- $Cn(A) \cap Cn(A') = Cn(\emptyset)$

A set $A \subseteq \mathcal{L}$ is *finitely representable* iff there is finite A' such that $Cn(A) = Cn(A')$.

closure under complement: A logic $\langle \mathcal{L}, Cn \rangle$ is closed under complement or simply *complemented* iff every finitely representable $A \subseteq \mathcal{L}$ has a complement $A' \subseteq \mathcal{L}$.

Dropping the restriction to finitely representable would narrow too much the scope of complemented logics. If this restriction was dropped, even CPL would not be complemented.[2]

Notice that the complement of a set $A \subseteq \mathcal{L}$ may not be unique. In the logic of Example 2.6 and Fig. 2.2 the set $\{a\}$ has two distinct complements $\{b\}$ and $\{c\}$.

---

[2] This was noted by Flouris in personal communication.

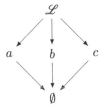

Fig. 2.2 Logic of Example 2.6

Example 2.6

$$\mathscr{L} = \{a, b, c\}$$
$$Cn(\{a, b\}) = Cn(\{b, c\}) = Cn(\{a, c\}) = \mathscr{L}$$
$$Cn(a) = \{a\}$$
$$Cn(b) = \{b\}$$
$$Cn(c) = \{c\}$$
$$Cn(\emptyset) = \emptyset$$

The following property guaranties the uniqueness of complement:

distribuitivity: A logic $\langle \mathscr{L}, Cn \rangle$ is distributive iff for all $A, B, C \in 2^{\mathscr{L}}$ we have:

$$Cn(A \cup B) \cap Cn(A \cup C) \subseteq Cn(A \cup (Cn(B) \cap Cn(C)))$$

By monotonicity we have that the converse of this property holds. Hence, Tarskian distributive logics satisfy the following property for every set A and every finitely representable sets B and C :

$$Cn(A \cup (Cn(B) \cap Cn(C))) = Cn(A \cup B) \cap Cn(A \cup C)$$

For now on when we mention distributive logics we mean Tarskian distributive logics.

A logic is called *Boolean* iff it is distributive and complemented. In Boolean logics, every sentence has an unique complement modulo equivalences:

Proposition 2.7 *Let* $\langle \mathscr{L}, Cn \rangle$ *be a Boolean logic and consider* $A \subseteq \mathscr{L}$. *If* A' *and* A'' *are two complements of* A *then* $Cn(A') = Cn(A'')$.

Proof

$$
\begin{aligned}
Cn(A') &= Cn(A' \cup Cn(\emptyset)) \\
&= Cn(A' \cup (Cn(A) \cap Cn(A''))) \\
&= Cn(A' \cup A) \cap Cn(A' \cup A'') \quad \text{by distribuitivity} \\
&= \mathscr{L} \cap Cn(A' \cup A'') \\
&= Cn(A' \cup A'')
\end{aligned}
$$

Using an analogous argument we prove that $Cn(A'') = Cn(A' \cup A'')$. Hence, $Cn(A') = Cn(A'')$. □

Furthermore, if two sets of sentences are equivalent in a Boolean logic then their complements are also equivalent:

Proposition 2.8 *Let* $\langle \mathscr{L}, Cn \rangle$ *be a Boolean logic and consider* $A, B \in 2^{\mathscr{L}}$ *and let* A' *be a complement of* A *and* B' *be a complement of* B *then* $Cn(A) = Cn(B)$ *if and only if* $Cn(A') = Cn(B')$.

Proof

$$
\begin{aligned}
Cn(A') &= Cn(A') \cup Cn(\emptyset) \\
&= Cn(A') \cup (Cn(B) \cap Cn(B')) \\
&= Cn(A') \cup (Cn(A) \cap Cn(B')) \\
&= Cn(A' \cup A) \cap Cn(A' \cup B') \qquad \text{by distribuitivity} \\
&= \mathscr{L} \cap Cn(A' \cup B') \\
&= Cn(A' \cup B')
\end{aligned}
$$

The proof that $Cn(B') = Cn(A' \cup B')$ is analogous and we conclude that $Cn(A') = Cn(B')$ □

In Boolean logics, the closure of the complement of A is denoted $\neg A$. $\neg A$ is well defined precisely because it is unique i.e., $\neg A = Cn(A')$ for A' a complement of A.

Let $A, K \in 2^{\mathscr{L}}$ such that $K = Cn(K)$ and $Cn(\emptyset) \subset A \subset K$, the *complement of A relative to* K (denoted $K^-(A)$) is the class of all sets K' such that:

- $Cn(K') \subset Cn(K)$
- $Cn(K' \cup A) = Cn(K)$

decomposability [FPA04]: A logic $\langle \mathscr{L}, Cn \rangle$ is *decomposable* iff for every $A, K \in 2^{\mathscr{L}}$ such that $K = Cn(K)$, $Cn(\emptyset) \subset Cn(A) \subset K$ and A is finitely representable we have that $K^-(A) \neq \emptyset$ i.e., there is $K' \subseteq \mathscr{L}$ such that $Cn(K') \subset Cn(K)$ and $Cn(K' \cup A) = Cn(K)$.

Decomposability was first introduced in [FPA04] and it is intimately related with AGM theory as will be showed in Chap. 4.

The following is a sufficient condition for a logic not to be decomposable:

Lemma 2.9 [FPA05] *Consider a logic* $\langle \mathscr{L}, Cn \rangle$. *If for some* $K, K' \in 2^{\mathscr{L}}$ *such that* $K' = Cn(\{\beta \in \mathscr{L} : Cn(\beta) \subset K\})$ *we have that* $Cn(\emptyset) \subset K' \subset Cn(K)$ *then* $\langle \mathscr{L}, Cn \rangle$ *is not decomposable.*

Proof For any $A \subseteq \mathscr{L}$, if $Cn(A) \subset Cn(K)$ then $Cn(A) \subseteq K'$. It follows that $Cn(K' \cup A) = Cn(K') \neq Cn(K)$. □

Now let us present some relations between decomposability and the other logical properties presented so far.

Proposition 2.10 *Boolean logics are decomposable.*

Proof Consider a Boolean logic $\langle \mathcal{L}, Cn \rangle$ and two sets of sentences $A, K \subseteq \mathcal{L}$ such that A is finitely representable and $Cn(\emptyset) \subset Cn(A) \subset Cn(K)$. We will prove that then $Cn(K) \cap Cn(A') \in K^-(A)$ for some complement A' of A (which must exist because $\langle \mathcal{L}, Cn \rangle$ is complemented).
Let $B = Cn(K) \cap Cn(A')$, then $Cn(A \cup B) = Cn(A \cup (Cn(K) \cap Cn(A')))$. By distributivity this is equal to $Cn(A \cup K) \cap Cn(A \cup A') = Cn(K)$.

If $A \subseteq B$ then $Cn(K) \cap Cn(A') = B = Cn(B) = Cn(B \cup A) = Cn(K)$, so $Cn(A') \supseteq Cn(K) \supset Cn(A)$. In this case, $Cn(A) = Cn(\emptyset)$ which is a contradiction. It follows that $A \nsubseteq B$. Hence, $B \in K^-(A)$.   □

Proposition 2.11 *If a logic is decomposable and satisfy the descending chain condition then it is complemented.*

Proof Let $A \subseteq \mathcal{L}$. By decomposability we have that $\mathcal{L}^-(A) \neq \emptyset$. By the descending chain condition, there is an $X' \in \mathcal{L}^-(A)$ such that $X' \subseteq X$ and there is no X'' such that $Cn(X'') \subset Cn(X')$. We will show that X' is a complement of A.

Since $X' \in \mathcal{L}^-(A)$ then $Cn(X' \cup A) = \mathcal{L}$. Now suppose that $Cn(X') \cap Cn(A) \neq Cn(\emptyset)$, by decomposability, there is $Y \in X'^-(Cn(X') \cap Cn(A))$.

In this case, we have that $X' = Cn(Y \cup (Cn(A) \cap Cn(X'))) \subseteq Cn(Y \cup A)$. However, since $A \subseteq Cn(Y \cup A)$, we have that $Cn(X' \cup A) \subseteq Cn(Y \cup A)$. Hence, $Cn(Y \cup A) = \mathcal{L}$.
In this case, $Cn(Y) \subset Cn(X')$ and $Cn(Y \cup A) = \mathcal{L}$. It follows that X' is not the minimal which contradicts the definition. We conclude that $Cn(X') \cap Cn(A) = Cn(\emptyset)$.
                                                                                         □

As a corollary of this result we have that finite decomposable logics are complemented.

2.3 Standard Languages

So far, no assumptions was made over the structure of the language $\mathcal{L}$. In this section, the standard language for propositional logics will be presented together with logical properties that depend on it.

A language $\mathcal{L}$ is *closed under an n-ary connective N* iff for every $\alpha_1, \ldots, \alpha_n \in \mathcal{L}$ we have that $N(\alpha_1, \ldots, \alpha_n) \in \mathcal{L}$. We will use standard infix notation for binary connectives i.e., we will write $\alpha N \beta$ instead of $N(\alpha, \beta)$.

A language is *standard* if it is closed under the standard connectives $\wedge$ (conjunction), $\vee$ (disjunction), $\rightarrow$ (implication) and $\neg$ (negation) (of course only the last connective is unary while the others are binary).

Consider a language $\mathcal{L}$ closed under negation. We say the negation $\neg$ in $\mathcal{L}$ is *classical* iff $\langle \mathcal{L}, Cn \rangle$ satisfies the following properties for every $\alpha \in \mathcal{L}$:

1. $Cn(\alpha) \cap Cn(\neg \alpha) = Cn(\emptyset)$
2. $Cn(\{\alpha, \neg \alpha\}) = \mathcal{L}$

If a set contains both α, $\neg\alpha \in Cn(A)$ then we say that A is *contradictory*. Using this terminology we say that the second statement above guarantees that if $\langle \mathscr{L}, Cn \rangle$ is closed under classical negation then if A is contradictory it must be trivial. Of course, if a logic is closed under classical negation then it is complemented.

The following property of negation states that the negation of a sentence never "helps" to prove the sentence [Was00]:

α-*local non contravention*: A logic $\langle \mathscr{L}, Cn \rangle$ closed under negation satisfies α-*local non contravention* iff for every $\alpha \in \mathscr{L}$ and every $A \subseteq \mathscr{L}$ we have that if $\alpha \notin Cn(A)$ then $\alpha \notin Cn(A \cup \{\neg\alpha\})$

Proposition 2.12 *If $\langle \mathscr{L}, Cn \rangle$ is distributive and closed under classical negation then $\langle \mathscr{L}, Cn \rangle$ satisfies α-local non contravention.*

Proof Let $\alpha \in Cn(A \cup \{\neg\alpha\})$. Since $\alpha \in Cn(A \cup \{\neg\alpha\}) \cap Cn(A \cup \{\alpha\})$, by distributivity, $\alpha \in Cn(A \cup (Cn(\alpha) \cap Cn(\neg\alpha)))$. Since $\neg$ is classical, $Cn(\alpha) \cap Cn(\neg\alpha) = Cn(\emptyset)$. It follows that $\alpha \in Cn(A)$. □

In Chap. 4 classical AGM theory will be presented. This theory makes certain assumptions about the underlying logic $\langle \mathscr{L}, Cn \rangle$. First, it assumes that the language is a standard language i.e., it is closed under the standard connectives. The other two assumptions are presented below:

deduction: A logic $\langle \mathscr{L}, Cn \rangle$ closed under implication satisfies *deduction* iff for every $\alpha \in \mathscr{L}$ and every $A \subseteq \mathscr{L}$ we have that $\alpha \in Cn(A \cup \{\beta\})$ iff $\beta \to \alpha \in Cn(A)$.
supraclassicality: A logic $\langle \mathscr{L}, Cn \rangle$ is *supraclassical* iff for every $\alpha \in \mathscr{L}$ and every $A \subseteq \mathscr{L}$ we have that if $\alpha \in C_{CPL}(A)$ then $\alpha \in Cn(A)$ i.e., if α is a classical consequence of A then $\alpha \in Cn(A)$. Classical consequence will be formalized in the next section.

A consequence relation R satisfies deduction if the following holds:

$$(A \cup \{\beta\}, \alpha) \in R \quad \text{iff} \quad (A, \beta \to \alpha) \in R$$

If R satisfies deduction then the consequence operator induced by R also satisfies deduction and conversely.

We say that a logic satisfies the *AGM-assumptions* if it satisfies four requirements: closure under standard language, deduction, supraclassicality, and compactness. We will sometimes call *well behaved* a logic that satisfies AGM assumptions.

Well-behaved logics are Boolean.

Lemma 2.13 *If $\langle \mathscr{L}, Cn \rangle$ satisfy the AGM-assumption then $Cn(A \cup \{\alpha_1\}) \cap Cn(A \cup \{\alpha_2\}) \subseteq Cn(A \cup \{\alpha_1 \vee \alpha_2\})$.*

Proof By deduction we have that $\alpha_1 \to \beta, \alpha_2 \to \beta \in Cn(A)$. Since $(\alpha_1 \vee \alpha_2) \to \beta \in C_{CPL}(\{\alpha_1 \to \beta, \alpha_2 \to \beta\})$, by supraclassicality, monotonicity and idempotence $(\alpha_1 \vee \alpha_2) \to \beta \in Cn(A)$. Finally, by deduction, $\beta \in Cn(A \cup \{\alpha_1 \vee \alpha_2\})$. □

Fig. 2.3 Relation between
logical properties

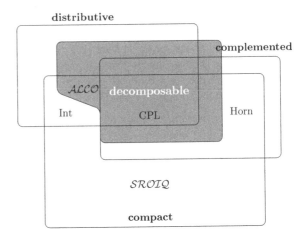

Proposition 2.14 *If a logic $\langle \mathscr{L}, Cn \rangle$ satisfies the AGM-assumptions then $\langle \mathscr{L}, Cn \rangle$ is Boolean.*

Proof Let $\beta \in Cn(A \cup B) \cap Cn(A \cup C)$. By compactness there are $A_1 \subseteq A \cup B$ and $A_2 \subseteq A \cup C$ both finite and such that $\beta \in Cn(A_1) \cap Cn(A_2)$. Let $A' = A_1 \cup A_2$, $B' = A_1 \cap A$, $C' = A_2 \cap C$. It is easy to verify that A', B' and C' are finite and that $\beta \in Cn(A' \cup B') \cap Cn(A' \cup C')$. Let $B' = \{\beta_0, \ldots, \beta_n\}$, $C' = \{\beta_{n+1}, \ldots \beta_{n+m}\}$, $\gamma_1 = \beta_0 \wedge \ldots \wedge \beta_n$ and $\gamma_2 = \beta_{n+1} \wedge \ldots \wedge \beta_{n+m}$. By supraclassicality we have that $Cn(A' \cup B') \cap Cn(A' \cup C') = Cn(A' \cup \{\gamma_1\}) \cap Cn(A' \cup \{\gamma_2\})$. Using Lemma 2.13 $Cn(A' \cup \{\gamma_1\}) \cap Cn(A' \cup \{\gamma_2\}) \subseteq Cn(A' \cup \{\gamma_1 \vee \gamma_2\})$. It follows that:

$$
\begin{aligned}
\beta \in\ & Cn(A' \cup B') \cap Cn(A' \cup C') \\
=\ & Cn(A' \cup \{\gamma_1\}) \cap Cn(A' \cup \{\gamma_2\}) \\
\subseteq\ & Cn(A' \cup \{\gamma_1 \vee \gamma_2\}) \\
\subseteq\ & Cn(A' \cup (Cn(B') \cap Cn(C'))) \\
\subseteq\ & Cn(A \cup (Cn(B) \cap Cn(C)))
\end{aligned}
$$

We conclude that $\beta \in Cn(A \cup (Cn(B) \cap Cn(C)))$ and, hence, $Cn(A \cup B) \cap Cn(A \cup C) \subseteq Cn(A \cup (Cn(B) \cap Cn(C)))$.

Now consider a set $X \subseteq \mathscr{L}$. If X is finitely representable then there is a finite X' such that $Cn(X) = Cn(X')$. Let $X' = \{\beta_0, \ldots, \beta_n\}$. Since $\langle \mathscr{L}, Cn \rangle$ is supraclassical we have that $Cn(\beta_0, \ldots, \beta_n) = Cn(\beta_0 \wedge \ldots \wedge \beta_n)$. We will show that $\neg\alpha$ is the complement of $\alpha = \beta_0 \wedge \ldots \wedge \beta_n$.

Since $\beta \in C_{CPL}(\alpha, \neg\alpha)$ (see Sect. 3.1) for every $\beta \in \mathscr{L}$ we have $\beta \in Cn(\alpha, \neg\alpha)$, by supraclassicality. Furthermore, let $\beta \in Cn(\alpha) \cap Cn(\neg\alpha)$. By deduction it holds that $\alpha \rightarrow \beta, \neg\alpha \rightarrow \beta \in Cn(\emptyset)$. Since $\alpha \vee \neg\alpha \in C_{CPL}(\emptyset)$ (see Sect. 3.1), by supraclassicality, we have that $\beta \in Cn(\emptyset)$. Hence, $Cn(\alpha) \cap Cn(\neg\alpha) \subseteq Cn(\emptyset)$. □

A trivial corollary of this result states that if a logic $\langle \mathscr{L}, Cn \rangle$ satisfies the AGM-assumptions then it is decomposable.

2.4 Conclusion

In this chapter, a very general framework for logics was presented. A logic was defined as a language together with a consequence operator. In this book, we are interested only in Tarskian logics i.e., in logics with a consequence operator that satisfies monotonicity, inclusion, and idempotence. Not every logic in the literature is Tarskian, non-monotonic logics [AW97] and linear logics [Gir87] are examples of non-Tarskian logics. These logics would not be considered in this book.

One nice property of Tarskian logics is that there is a bijection between them and complete lattices. For this reason, it is possible to visually illustrate certain toy logics. We will use lattice many times throughout the book as an heuristic to generate examples of logics with certain properties. For a more complete presentation of consequence operator see [Wój88].

Tarskian logics may satisfy certain properties such as distributivity, decomposability etc. These properties are not independent. Some combinations of properties may imply other properties. Several such relations between properties were proved in this chapter. Diagram in Fig. 2.3 sums up the these results. The points in the diagram correspond to logics that will be presented in the following chapter.

References

[AW97]   G. Antoniou and M.A. Williams. *Nonmonotonic reasoning*. Artificial intelligence. MIT Press, 1997.

[Flo06]   Giorgos Flouris. *On Belief Change and Ontology Evolution*. PhD thesis, University of Crete, 2006.

[FPA04]   Giorgos Flouris, Dimitris Plexousakis, and Grigoris Antoniou. Generalizing the AGM postulates: preliminary results and applications. In James P. Delgrande and Torsten Schaub, editors, *Proceedings of the 10th International Workshop on Non-Monotonic Reasoning 2004 (NMR-04)*, pages 171–179, Whistler BC, Canada, June 6-8 2004.

[FPA05]   Giorgos Flouris, Dimitris Plexousakis, and Grigoris Antoniou. On applying the AGM theory to DLs and OWL. In Enrico Motta Yolanda Gil, V. Richard Benjamins, and Mark A. Musen, editors, *Proceedings of the 4th International Semantic Web Conference (ISWC 2005)*, pages 216–231, Galway, Ireland, November, 6-10 2005. Springer.

[Gir87]   Jean-Yves Girard. Linear logic. *Theoretical Computer Science*, 50:1–102, 1987.

[Was00]   Renata Wassermann. *Resource Bounded Belief Revision*. PhD thesis, Universiteit van Amsterdam, Janeiro 2000.

[Wój88]   R. Wójcicki. *Theory of logical calculi: basic theory of consequence operations*. Synthese library. Kluwer Academic Publishers, 1988.

Chapter 3
Logics

Abstract In this chapter, a list of Tarskian logics is presented, namely, classical propositional logic (CPL), intuitionistic logic (Int), horn logic (Horn), and description logics (DLs). Classical Propositional Logic is the canonical example of well-behaved logic logic, i.e., a logic that satisfies the AGM assumptions. Besides the interest in the properties that these logics satisfy, they were chosen for diverse reasons. Intuitionistic logic has great interest among logicians as the basis for constructive mathematics. Horn logic and Description Logics have great interest in computer science because of their computational complexity. The interest in Description Logics in particular has grown quickly among computer scientist since the announcement of a DL-based language, OWL, as the standard language to describe ontologies on the Web.

Keywords Non-classical logics · Horn logic · Description Logics · OWL · Intuitionistic logic

We present classical propositional logic (CPL) in Sect. 3.1 as an example of logic where the classical AGM theory is applicable. Intuitionistic logic is presented in Sect. 3.2, Horn logic in Sect. 3.3, and Description Logics (DLS) in Sect. 3.4. As will be shown in further chapters, Int and Horn as well as many DLs are examples of logics where classical AGM theory is *not* applicable. These later logics (Intuitionistic, Horn, and Description logics) satisfy interesting properties that will be explored in further chapters.

Lots of other logics could be added to this list. Logics that are not Tarskian, such as non-monotonic logics [AW97] and linear logics [Gir87], and logics that are not compact, such as Computational Tree Logic (CTL) [EH86] were discarded because they do not fit the framework presented in this book. Other logics such as modal logics [Che80] were not listed because, for our purposes, they have properties similar to other logics listed here.

M. M. Ribeiro, *Belief Revision in Non-Classical Logics*,
SpringerBriefs in Computer Science,
DOI: 10.1007/978-1-4471-4186-0_3, © The Author(s) 2013

3.1 Classical Propositional Logic

In this section a brief introduction to CPL is presented. First, it is presented the language of CPL. After that the semantic consequence relation and the syntactic consequence relation for CPL are defined. In the end of the section some properties of CPL are proved.

This chapter does not intend to be a complete introduction to CPL, it just presents some basic definitions enough to prove the properties needed for the rest of this book.

3.1.1 Language

Let $\mathbb{P} = \{p_0, p_1, \dots\}$ be an infinite enumerable set called the *signature* of CPL. The elements of $\mathbb{P}$ are called *propositional variables*. The language of CPL is the least set $\mathscr{L}_{CPL}$ such that:

1. $\mathbb{P} \subseteq \mathscr{L}_{CPL}$
2. if $\alpha, \beta \in \mathscr{L}_{CPL}$ then $(\alpha \wedge \beta), (\alpha \vee \beta), (\alpha \rightarrow \beta), (\neg\alpha) \in \mathscr{L}_{CPL}$.

Another way of presenting a language is using Backus-Naur form (BNF):

$$\alpha := p_i | (\alpha \vee \alpha) | (\alpha \wedge \alpha) | (\alpha \rightarrow \alpha) | (\neg\alpha)$$

Example 3.1. Let $p_0, p_1, \dots \in \mathbb{P}$.

$$p_0 \in \mathscr{L}_{CPL}$$
$$(\neg p_0) \in \mathscr{L}_{CPL}$$
$$(\neg(\neg p_0)) \in \mathscr{L}_{CPL}$$
$$(p_1 \rightarrow (\neg(\neg p_0))) \in \mathscr{L}_{CPL}$$
$$(p_0\neg \rightarrow p_1) \notin \mathscr{L}_{CPL}$$

The presentation using BNF is convenient for more complex languages. For this reason it will be used to present other languages in this book. Standard conventions for the parenthesis omission will be used for this and the rest of the languages presented in this book.

Notice that $\mathscr{L}_{CPL}$ is a standard language. In fact, it is the least standard language that contains $\mathbb{P}$.

3.1.2 Semantic

A function $v : \mathbb{P} \rightarrow \{0, 1\}$ is called a *truth-function*. A truth-function assigns the value 0 (false) or 1 (true) for every propositional variable. A truth-function *satisfies* (or *is a model for*) a sentence α written $\models_v \alpha$ iff:

- $\models_v p$ iff $v(p) = 1$.
- $\models_v \neg\alpha$ iff it is not the case that $\models \alpha$ (in symbols $\not\models_v \alpha$).
- $\models_v \alpha \vee \beta$ iff $\models_v \alpha$ or $\models_v \beta$.
- $\models_v \alpha \wedge \beta$ iff $\models_v \alpha$ and $\models_v \beta$.
- $\models_v \alpha \rightarrow \beta$ iff $\not\models_v \alpha$ or $\models_v \beta$.

A truth-function v satisfies a set A iff it satisfies every sentence $\beta \in A$, i.e., $\models_v A$ iff $\models_v \beta$ for every $\beta \in A$. A sentence α is a *semantic consequence* of a set of sentences A (written $A \models_{CPL} \alpha$) iff for every truth-function v that satisfies A, v also satisfies α, i.e., if $\models_v A$ implies $\models_v \alpha$. From now on $\models$ will be used instead of $\models_{CPL}$ whenever it is clear from the context that we are dealing with CPL.

Example 3.2 If $v(p) = 1$ then $\models_v p$ and in this case $\models_v p \vee \neg p$.
  On the other hand if $v(p) = 0$ then $\models_v \neg p$ and $\models_v p \vee \neg p$. Hence, for all truth-function v we have that $\models_v p \vee \neg p$, i.e., $\models p \vee \neg p$.

Example 3.3 If $\models_v \neg\neg p$ then $\not\models_v \neg p$ and it follows that $\models_v p$. Hence, $\neg\neg p \models p$.

It is easy to verify that the relation $\models$ is Tarskian.

3.1.3 Axiomatic System

Consider an enumerable set Ξ whose elements are called *schema variables*. Let $\mathscr{L}_\Xi$ be the least standard language that contains Ξ. A *rule* is a pair $\frac{A}{\alpha}$ such that $A \subseteq \mathscr{L}_\Xi$ is the set of *premises* of the rule and $\alpha \in \mathscr{L}_\Xi$ is the *consequence* of the rule. An *axiom schema* is a rule with an empty set of premises. To simplify notation, an axiom schema is represented simply as a sentence $\alpha \in \mathscr{L}_\Xi$ rather than $\frac{\emptyset}{\alpha}$. An *axiomatic system* is just a set of rules.

A *substitution* is a function $\delta : \Xi \rightarrow \mathscr{L}_\Xi$ from schema variables to sentences. A substitution can be extended uniquely to a function $\hat{\delta} : \mathscr{L}_\Xi \rightarrow \mathscr{L}_\Xi$ over the set of formulas in the standard following way:

- $\hat{\delta}(c) = c$ if $c \in \mathbb{P}$
- $\hat{\delta}(\neg\alpha) = \neg\hat{\delta}(\alpha)$
- $\hat{\delta}(\alpha_1 \wedge \alpha_2) = \hat{\delta}(\alpha_1) \wedge \hat{\delta}(\alpha_2)$
- $\hat{\delta}(\alpha_1 \vee \alpha_2) = \hat{\delta}(\alpha_1) \vee \hat{\delta}(\alpha_2)$
- $\hat{\delta}(\alpha_1 \rightarrow \alpha_2) = \hat{\delta}(\alpha_1) \rightarrow \hat{\delta}(\alpha_2)$

Given an axiomatic system Γ, a *derivation* from $A \subseteq \mathscr{L}_{CPL}$ to $\alpha \in \mathscr{L}_{CPL}$ (written $A \vdash_\Gamma \alpha$) is a sequence $\alpha_0, \ldots \alpha_n$ such that $\alpha_n = \alpha$, $\alpha_i \in \mathscr{L}$ is an element of A or there is a substitution σ and an inference rule $\frac{B}{\beta} \in \Gamma$ such that $\sigma(\beta) = \alpha_i$ and $\sigma(\beta') = \alpha_j$ for every $\beta' \in B$ and $j < i$.
  Consider the following axiomatic system:

modus ponens: $\frac{\{\xi_1 \rightarrow \xi_2, \xi_1\}}{\xi_2}$
axiom 1: $\xi_1 \rightarrow (\xi_2 \rightarrow \xi_1)$

axiom 2: $(\xi_1 \rightarrow (\xi_2 \rightarrow \xi_3)) \rightarrow ((\xi_1 \rightarrow \xi_2) \rightarrow (\xi_1 \rightarrow \xi_3))$
axiom 3: $\xi_1 \rightarrow (\xi_2 \rightarrow (\xi_1 \wedge \xi_2))$
axiom 4: $(\xi_1 \wedge \xi_2) \rightarrow \xi_1$
axiom 5: $(\xi_1 \wedge \xi_2) \rightarrow \xi_2$
axiom 6: $\xi_1 \rightarrow (\xi_1 \vee \xi_2)$
axiom 7: $\xi_2 \rightarrow (\xi_1 \vee \xi_2)$
axiom 8: $\neg\xi_1 \rightarrow (\xi_1 \rightarrow \xi_2)$
axiom 9: $(\xi_1 \rightarrow \xi_2) \rightarrow ((\neg\xi_1 \rightarrow \xi_2) \rightarrow \xi_2)$
axiom 10: $((\xi_1 \vee \xi_2) \wedge \neg\xi_1) \rightarrow \xi_2$

In further sections we might write $\vdash_{CPL}$ to distinguish this axiomatic system from the others that may appear. Whenever it is clear by the context, though, we will just write $\vdash$.

Example 3.4 The following sequence is a derivation of $\vdash \alpha \rightarrow \alpha$:

$\alpha \rightarrow ((\alpha \rightarrow \alpha) \rightarrow \alpha)$	ax. 1
$(\alpha \rightarrow ((\alpha \rightarrow \alpha) \rightarrow \alpha)) \rightarrow ((\alpha \rightarrow (\alpha \rightarrow \alpha)) \rightarrow (\alpha \rightarrow \alpha))$	ax. 2
$(\alpha \rightarrow (\alpha \rightarrow \alpha)) \rightarrow (\alpha \rightarrow \alpha)$	m.p. 1,2
$\alpha \rightarrow (\alpha \rightarrow \alpha)$	ax. 2
$\alpha \rightarrow \alpha$	m.p. 3,4

We use ax. n before a sentence α to indicate that α was derived by a substitution of the axiom n. The symbols m.p. i, j are used before a sentence to indicate that this sentence was derived by modus ponens with the sentences i and j as premises.

Example 3.5 The following is derivation for $\{\alpha \vee \beta, \neg\alpha\} \vdash \beta$.

$\alpha \vee \beta$	hyp.
$(\alpha \vee \beta) \rightarrow (\neg\alpha \rightarrow (\neg\alpha \wedge (\alpha \vee \beta)))$	ax. 3
$\neg\alpha \rightarrow (\neg\alpha \wedge (\alpha \vee \beta))$	m.p. 1, 2
$\neg\alpha$	hyp.
$\neg\alpha \wedge (\alpha \vee \beta)$	m.p. 3, 4
$((\alpha \vee \beta) \wedge \neg\alpha) \rightarrow \beta$	ax. 10
β	m.p. 5, 6

Notice that hyp. was used before a sentence α to indicate that α was in the set of premises.

If every rule in $\frac{A}{\alpha}$ in the axiom system is *finitary* (i.e. if A is finite) then the logic $\langle \mathcal{L}, Cn \rangle$ induced by this consequence relation is trivially compact.

Proposition 3.6 *The relation $\vdash$ is Tarskian.*

Proof The first and third items of the definition of Tarskian relation are trivial to verify. For the second, notice that if $A \vdash \alpha$ then by compactness mentioned above there is a finite $A' \subseteq A$ such that $A' \vdash \alpha$. To conclude the proof notice that the concatenation of the deductions from B to the elements of A' with the derivation from A' to α is a derivation from B to α. □

Theorem 3.7 (Deduction). *The relation $\vdash$ satisfies deduction.*

Proof If $A \vdash \beta \to \alpha$ then by modus ponens we have that $A \cup \{\beta\} \vdash \alpha$. Conversely, if $A \vdash \beta$ then there is a derivation $\beta_0, \ldots, \beta_n$ with $\beta_n = \alpha$. Let $n = 0$, then $\beta_0 = \alpha$ and either α is an instance of one of the axioms or $\alpha \in A \cup \{\beta\}$.

If α is an axiom or if $\alpha \in A$ then the $\beta \to \alpha$ by axiom 1. Otherwise $\alpha = \beta$ and in this case, by example 3.4, $A \vdash \alpha \to \alpha$.

Now assume that for all n if there is a derivation $\beta_0, \ldots, \beta_n$ from $A \cup \{\beta\}$ to α then $A \vdash \beta \to \alpha$.

Now let $\beta_0, \ldots, \beta_{n+1}$ be a derivation from $A \cup \{\beta\}$ to α. We have three options: (1) α is an instance of an axiom, (2) $\alpha \in A \cup \{\beta\}$ or (3) there are $i, j \leq n$ such that $\beta_j = \beta_i \to \alpha$. Case (1) and (2) were covered in the base case. For case (3) we have by induction that $A \vdash \beta \to \beta_i$ and $A \vdash \beta \to (\beta_i \to \alpha)$. Hence, using axiom 2 we have $A \vdash \beta \to \alpha$. □

A set A is called *inconsistent* iff $A \vdash \alpha \wedge \neg\alpha$ for some $\alpha \in \mathscr{L}_{\mathrm{CPL}}$.

Proposition 3.8 (Triviality). *If A is inconsistent then A is trivial, i.e., $A \vdash \beta$ for every $\beta \in \mathscr{L}_{\mathrm{CPL}}$.*

Proof

$$
\begin{array}{ll}
\neg\alpha & \text{hyp} \\
\neg\alpha \to (\alpha \to \beta) & \text{ax. 8} \\
\alpha \to \beta & \text{m.p. 1,2} \\
\alpha & \text{hyp.} \\
\beta & \text{m.p. 3,4}
\end{array}
$$

Hence, for any $\beta \in \mathscr{L}_{\mathrm{CPL}}$ were have that $A \vdash \beta$. □

There is a close relation between the semantic consequence presented in the beginning of the section and the syntactic consequence presented above. First it is true that α is a syntactic consequence of A then it is also a semantic consequence of it, i.e., $A \vdash \alpha$ implies $A \vDash \alpha$. In other words, $\vdash$ is sound w.r.t. $\vDash$.

Theorem 3.9 (Soundness) *If $A \vdash \alpha$ then $A \vDash \alpha$.*

Proof First we need to check that for each rule $\frac{A}{\alpha}$ in Γ it holds that if $\vDash \sigma(A)$ then $\vDash \sigma(\alpha)$. Since this is very simple and tedious, we will show only two examples: the first axiom and modus ponens:

Let $\sigma(\xi_1 \to (\xi_2 \to \xi_1)) = \alpha \to (\beta \to \alpha)$ for $\alpha, \beta \in \mathscr{L}_{\mathrm{CPL}}$. Assume that $v(\alpha) = 0$ then $v(\alpha \to (\beta \to \alpha)) = 1$. Assume then that $v(\alpha) = 1$ then $v(\beta \to \alpha) = 1$, and hence $v(\alpha \to (\beta \to \alpha)) = 1$. We conclude that for any truth-function v we have that $v(\alpha \to (\beta \to \alpha)) = 1$.

Now let $\alpha_1 = \sigma(\xi_1)$ and $\alpha_2 = \sigma(\xi_2)$. If $v(\alpha_1) = v(\alpha_1 \to \alpha_2) = 1$ then, by definition, $v(\alpha_2) = 1$. □

Besides $\vdash$ being sound it is also complete w.r.t. $\vDash$, i.e., $A \vDash \alpha$ implies that $A \vdash \alpha$.

Lemma 3.10 (Lindenbaum) *If A is consistent then it can be extended to a consistent complete consistent set X, i.e., a consistent set such that for all $\beta \in \mathscr{L}_{CPL}$ either $\beta \in X$ or $\neg\beta \in X$.*

Proof Arrange the sentences of $\mathscr{L}_{CPL}$ is a sequence $\beta_1, \beta_2 \dots$. Now let $X_0 = A$ and for $i > 0$ let:

$$X_i = \begin{cases} X_{i-1} \cup \{\beta_i\} & \text{if consistent} \\ X_{i-1} \cup \{\neg\beta_i\} & \text{otherwise.} \end{cases}$$

Suppose that both $X_i \cup \{\beta\}$ and $X_i \cup \{\neg\beta\}$ are inconsistent. By triviality and deduction theorem $X_i \vdash \alpha \to \beta$ and $X_i \vdash \neg\alpha \to \beta$. Using axiom 9 we have $X_i \vdash \beta$. Using a similar argument $X_i \vdash \neg\beta$. Hence X_i is inconsistent which contradicts the hypothesis. We conclude that X_i is consistent for every i.

Now let $X = \bigcup_i X_i$. Suppose that X is inconsistent. By compactness there are finite sets $X' \subseteq X$ and $X'' \subseteq X$ such that $X' \vdash \beta$ and $X'' \vdash \neg\beta$ for some β. It follows that $X' \cup X''$ is finite and inconsistent. Since $X' \cup X'' \subseteq X$ then there is an i such that $X' \cup X'' = X_i$. This contradicts the fact that every X_i is consistent. Hence $X = \bigcup X_i$ is consistent.

To finish, notice also that, by construction, X is complete.   □

Theorem 3.11 (Completeness) *If $A \vDash \alpha$ then $A \vdash \alpha$.*

Proof We will show the contrapositive, i.e., if $A \nvdash \alpha$ then $A \nvDash \alpha$.

If $A \nvdash \alpha$ then $A \cup \{\neg\alpha\}$ is consistent and, by Lindenbaum lemma, it can be extended to a complete consistent set X.

Consider now a truth-function v such that $v(\beta) = 1$ if $\beta \in X$ and $v(\beta) = 0$ otherwise. We need to show that v is indeed a truth-function. First notice that since X is maximal the $X \vdash \beta$ iff $\beta \in X$.

($\neg$) $v(\neg\beta_1) = 1$ iff $\neg\beta \in X$ iff $\beta \notin X$ iff $v(\beta) = 0$

($\to$) $v(\beta_1 \to \beta_2) = 1$ iff $\beta_1 \to \beta_2 \in X$. Assume that $\beta_1 \to \beta_2 \in X$ and $\beta_1 \in X$ then by modus ponens $\beta_2 \in X$. It follows that $\beta_1 \notin X$ or $\beta_2 \in X$. Assume now that $\beta_1 \notin X$ or $\beta_2 \in X$. In the first case $\beta_1 \to \beta_2 \in X$ by axiom 8 and in the second $\beta_1 \to \beta_2 \in X$ by axiom 1.

It follows that $\beta_1 \to \beta_2 \in X$ iff $\beta_1 \notin X$ or $\beta_2 \in X$. Hence $v(\beta_1) = 0$ or $v(\beta_2) = 1$.

($\wedge$) $v(\beta_1 \wedge \beta_2) = 1$ iff $\beta_1 \wedge \beta_2 \in X$. Assume that $\beta_1 \wedge \beta_2 \in X$, then by axioms 4 and 5 $\beta_1, \beta_2 \in X$.

Assume now that $\beta_1, \beta_2 \in X$ then axiom 3 $\beta_1 \wedge \beta_2 \in X$.

It follows that $\beta_1 \wedge \beta_2 \in X$ iff $\beta_1, \beta_2 \in X$ iff $v(\beta_1) = v(\beta_2) = 1$.

($\vee$) $v(\beta_1 \vee \beta_2) = 1$ iff $\beta_1 \vee \beta_2 \in X$.

Assume first that $\beta_1 \in X$ or $\beta_2 \in X$. In both cases, by axioms 6 and 7, $\beta_1 \vee \beta_2 \in X$.

Now assume that $\beta_1 \vee \beta_2 \in X$. If $\beta_1 \notin X$ then by Example 3.5 $\beta_2 \in X$.

It follows that $\beta_1 \vee \beta_2 \in X$ iff $\beta_1 \in X$ or $\beta_2 \in X$ iff $v(\beta_1) = 1$ or $v(\beta_2) = 0$.

To conclude the proof notice that the truth-function v constructed above satisfies $v(A) = 1$ and $v(\alpha) = 0$. Hence $A \nvDash \alpha$.   □

3.1.4 Properties

By Theorems 3.9 and 3.11 we have that the consequence operator C_{CPL} induced either by the syntactic consequence relation $\vdash_{CPL}$ or by the semantic consequence relation $\vDash_{CPL}$ are equal, i.e.,

$$C_{CPL}(A) = \{\alpha \in \mathscr{L} : A \vdash_{CPL} \alpha\} = \{\alpha \in \mathscr{L} : A \vDash_{CPL} \alpha\}$$

In this section some properties of this operator will be presented.

Proposition 3.12 $\langle \mathscr{L}_{CPL}, C_{CPL} \rangle$ *satisfies the AGM assumptions.*

Proof Of course the language $\mathscr{L}_{CPL}$ is standard and of course C_{CPL} is supra-classical. C_{CPL} is compact, since it is induced by $\vdash$ which is compact and satisfies induction. Finally, deduction follows from Theorem 3.7.   □

As a corollary of the above proposition we have the following:

Proposition 3.13 $\langle \mathscr{L}_{CPL}, C_{CPL} \rangle$ *is Boolean and, hence, decomposable.*

Proof Follows directly form Propositions 3.12, 2.14 and 2.10.   □

Furthermore, the negation $\neg$ of CPL is classical.

Proposition 3.14 *For every* $\alpha \in \mathscr{L}_{CPL}$ *it holds that*:

- $C_{CPL}(\{\alpha, \neg\alpha\}) = \mathscr{L}$ *and*
- $C_{CPL}(\alpha) \cap Cn(\neg\alpha) = Cn(\emptyset)$.

Proof The first item follows from Proposition 3.8. For the second item assume that $\beta \in C_{CPL}(\alpha) \cap C_{CPL}(\neg\alpha)$. By deduction $\alpha \to \beta, \neg\alpha \to \beta \in C_{CPL}(\emptyset)$. Using Example 3.5 $\beta \in C_{CPL}(\emptyset)$. It follows that $C_{CPL}(\alpha) \cap C_{CPL}(\neg\alpha) \subseteq C_{CPL}(\emptyset)$. The rest of the proof is trivial.   □

As a corollary we have that in CPL the negation satisfies α-local non contravention.

CPL was presented as a canonical example of well-behaved logic and, hence, the canonical example of logic where classical belief revision is applicable (Chap. 4). It satisfies all AGM assumptions and, hence, it is Boolean and decomposable. Furthermore, the negation in CPL is classical.

3.2 Intuitionistic Logic

Intuitionistic logic is a formal system that avoids non-constructive proofs allowed in classical logic. Consider the following informal proof of the existence of two irrational numbers x and y such that x^y is rational [CE06].

Example 3.15 If $\sqrt{2}^{\sqrt{2}}$ was rational then we are done. Suppose then that $\sqrt{2}^{\sqrt{2}}$ is irrational. Since $(\sqrt{2}^{\sqrt{2}})^{\sqrt{2}} = 2$ is rational and $\sqrt{2}$ is irrational, it follows that there are x and y irrationals such that x^y is rational.

Some mathematicians do not accept this kind of proof. The problem they point in the prove above is that, it does not present the numbers x and y. In other words, this proof is not constructive.

Intuitionistic logic was developed to avoid this kind of argument. In this section we will present only the propositional version of intuitionistic logic. For a complete first-order presentation see [Min00].

The language $\mathscr{L}_{Int}$ of Intuitionistic Logic is exactly the same of Classical Propositional Logic, i.e., $\mathscr{L}_{Int} = \mathscr{L}_{CPL}$. The consequence operator C_{Int}, however, must be different to avoid deductions similar to the one in example 3.15. Notice that this proof deeply depends on the rule of excluded middle: $\xi \vee \neg\xi$. Hence, Intuitionistic Logic reject this principle.

In what follows a Kripke semantic and an axiomatic system for Int will be presented. The section ends showing some properties of $\langle \mathscr{L}_{Int}, C_{Int} \rangle$.

3.2.1 Semantics

Consider an enumerable set $\mathbb{P} = \{p_0, p_1 \ldots\}$ of propositional variables. A *Kripke model* for intuitionistic logic is a triple $M = \langle W, \leq, v \rangle$ where W is any enumerable set whose elements are called *worlds*, $\leq$ is a partial order over W called the *accessibility relation* and $v : \mathbb{P} \to 2^W$ a function that associates propositional variables to sets of worlds.

Given a Kripke model M, we define the relation $\vDash_M \subseteq W \times \mathscr{L}_{Int}$ as follows:

- $w \vDash_M p$ iff $w \in v(p)$.
- $w \vDash_M \alpha \vee \beta$ iff $w \vDash_v \alpha$ or $w \vDash_v \beta$.
- $w \vDash_M \alpha \wedge \beta$ iff $w \vDash_v \alpha$ and $\vDash_v \beta$.
- $w \vDash_M \neg\alpha$ iff $w' \nvDash_v \alpha$ for every $w \leq w'$.
- $w \vDash_M \alpha \to \beta$ iff $w' \nvDash_v \beta$ for all $w \leq w'$ such that $w' \vDash_v \alpha$.

This relation intuitively relates worlds to sentences valid in those worlds.

We will write $w \vDash A$ for $A \subseteq \mathscr{L}_{Int}$ iff $w \vDash \beta$ for every $\beta \in A$, i.e., a set A is valid in a world w iff each sentence of A is valid in w. A sentence $\alpha \in \mathscr{L}_{Int}$ is a *semantic consequence* of $A \subseteq \mathscr{L}_{\text{Int}}$, written $A \vDash_{\text{Int}} \alpha$, iff for every world $w \in W$ we have that if $w \vDash A$ then $w \vDash \alpha$. Again, $\vDash$ will be used instead of $\vDash_{\text{Int}}$ when the context is clear.

Example 3.16 Let us show that $\nvDash p \vee \neg p$.
Consider the following model $M = \langle W, \leq, v \rangle$:

$$
\begin{aligned}
W &= \{w_1, w_2\} \\
\leq &= \{(w_1, w_2)\} \\
v(p) &= \{w_2\}
\end{aligned}
$$

Since $w_1 \notin v(p)$, we have that $w_1 \nvDash_M p$. Furthermore, $w_1 \nvDash_M \neg p$ since $w_1 \leq w_2$ and $w_2 \vDash_M \neg p$. Hence, $\nvDash p \vee \neg p$.

Example 3.17 Now we show that $\neg\neg p \nvDash p$.
Consider the model M in the above Example. As already argued $w_1 \nvDash_M p$. Furthermore, $w_2 \vDash p$ and then $w_1 \nvDash \neg p$. It follows that $w_1 \vDash \neg\neg p$. Hence, $\neg\neg p \nvDash p$.

Notice that the relation $\vDash_{\text{Int}}$ is Tarskian and not supraclassical, since $p \vee \neg p$ is a theorem of CPL.

3.2.2 Axiomatic System

Consider the following axiomatic system:

modus ponens: $\dfrac{\{\xi_1 \to \xi_2, \xi_1\}}{\xi_2}$
axiom 1: $\xi_1 \to (\xi_2 \to \xi_1)$
axiom 2: $(\xi_1 \to \xi_2) \to ((\xi_1 \to (\xi_2 \to \xi_3)) \to (\xi_1 \to \xi_3))$
axiom 3: $\xi_1 \to (\xi_2 \to (\xi_1 \wedge \xi_2))$
axiom 4: $(\xi_1 \wedge \xi_2) \to \xi_1$
axiom 5: $(\xi_1 \wedge \xi_2) \to \xi_2$
axiom 6: $\xi_1 \to (\xi_1 \vee \xi_2)$
axiom 7: $\xi_2 \to (\xi_1 \vee \xi_2)$
axiom 8: $\neg\xi_1 \to (\xi_1 \to \xi_2)$
axiom 9: $(\xi_1 \to \xi_2) \to ((\xi_1 \to \neg\xi_2) \to \neg\xi_1)$
axiom 10: $(\xi_1 \vee \xi_2) \to ((\xi_1 \to \xi_3) \to ((\xi_2 \to \xi_3) \to \xi_3))$

First notice that axioms 1 to 8 coincide with the rules of the axiomatic system presented for CPL. Of course if $A \vdash_{\Gamma'} \alpha$ and $\Gamma' \subseteq \Gamma$ then $A \vdash_\Gamma \alpha$. Hence, some results from Sect.3.1 hold in this axiomatic system, e.g., Proposition 3.8 and Theorem 3.7.

This axiomatic system is compact, since it does not have any infinitary rule. Using the same argument presented in previous section we have that $\vdash$ is Tarskian. Furthermore, $\vdash_{\text{Int}}$ is sound w.r.t. $\vDash_{\text{Int}}$.

Theorem 3.18 (Soundness) *If $A \vdash \alpha$ then $A \vDash \alpha$.*

Proof This proof follows the same lines as the one in Theorem 3.9 and will be omitted here.   □

Now for the completeness we need some definitions and lemmas. We say that a set $X \subseteq \mathscr{L}_{\text{Int}}$ is *full* iff it satisfies the following:

1. X is consistent.
2. If $X \vdash \alpha$ then $\alpha \in X$.
3. If $\alpha_1 \vee \alpha_2 \in X$ then $\alpha_1 \in X$ or $\alpha_2 \in X$.

If $B \subseteq B'$ and B' is full then B' is called a *full extension* of B.

Lemma 3.19. *If $X \nvdash \alpha$ then there is a full extension X' of X such that $\alpha \notin X'$*

Proof Arrange the elements of $\mathscr{L}_{\text{Int}}$ in a sequence $\beta_1, \beta_2 \ldots$. Let $X_0 = X$ and for $i > 0$ let:

$$X_i = \begin{cases} X_{i-1} \cup \{\beta_i\} & \text{if } X_{i-1} \nvdash \beta_i \to \alpha \\ X_{i-1} & \text{otherwise.} \end{cases}$$

Finally, let $X' = \bigcup X_i$.

First let using prove that $X' \nvdash \alpha$. By construction $X_0 \nvdash \alpha$ and for $i > 0$ if $X_{i-1} \cup \{\beta_i\} = X_i \vdash \alpha$ then by deduction theorem (Theorem 3.7) $X_{i+1} \vdash \beta_i \to \alpha$ which is a contradiction. It follows that no finite subset of X' derives α and by compactness $X \nvdash \alpha$.

Now we have to prove that X' is full. By Proposition 3.8 X' is consistent, since $X' \nvdash \alpha$. Now suppose that $X' \vdash \beta$, but $\beta \notin X'$. By construction $X' \vdash \beta \to \alpha$ and by modus ponens $X' \vdash \alpha$ which is a contradiction. Finally, suppose that $\beta_1 \vee \beta_2 \in X'$, $\beta_1 \notin X'$ and $\beta_2 \notin X'$. By construction $\beta_1 \to \alpha, \beta_2 \to \alpha \in X'$ and using axiom 10 we have $X \vdash \alpha$ which is a contradiction.   □

Theorem 3.20 (Completeness) *If $A \vDash \alpha$ then $A \vdash \alpha$.*

Proof Consider the following Kripke model $M = \langle W, \leq, v \rangle$:

$$W = \{B \subseteq \mathscr{L} : B \text{ is full and } A \subseteq B\}$$
$$\leq\, = \{(B_1, B_2) \in 2^{\mathscr{L}_{\text{Int}}} \times 2^{\mathscr{L}_{\text{Int}}} : B_1 \subseteq B_2\}$$
$$v(p) = \{B \subseteq \mathscr{L} : p \in B\}$$

By Lemma 3.19 $W \neq \emptyset$. We will show that for any $B \subseteq \mathscr{L}$ and any $\beta \in \mathscr{L}$, $B \vDash_M \beta$ iff $\beta \in B$.

This is proved inductively in the construction of the formulas in $\mathscr{L}_{\text{Int}}$. For the base $\beta = p \in \mathbb{P}$ and it follows $B \vDash_M p$ iff $p \in B$ by construction.

($\wedge$) $B \vDash_M \beta_1 \wedge \beta_2$ iff $B \vDash_M \beta_1$ and $B \vDash_M \beta_2$ iff, by induction hypothesis, $\beta_1 \in B$ and $\beta_2 \in B$. It follows that $\beta_2 \in B$ then using axiom 3 $\beta_1 \wedge \beta_2 \in B$.
Conversely, if $\beta_1 \wedge \beta_2 \in B$ then by axiom 4 and 5 $\beta_1, \beta_2 \in B$.

($\vee$) $B \vDash_M \beta_1 \vee \beta_2$ iff $B \vDash_M \beta_1$ or $B \vDash_M \beta_2$ iff, by induction hypothesis, $\beta_1 \in B$ or $\beta_2 \in B$. It follows, by axioms 6 and 7, that $\beta_1 \vee \beta_2 \in B$.

Conversely, if $\beta_1 \vee \beta_2 \in B$ then $\beta_1 \in B$ or $\beta_2 \in B$, since B is full.

($\neg$) $B \vDash_M \neg\beta$ iff, by induction hypothesis, $\beta \notin B'$ for all full extension B' of B. Notice that $B \cup \{\beta\}$ must be inconsistent, otherwise, by Lemma 3.19 there would be a full extension B' of B which included β. Hence, by Proposition 3.8 $B \cup \{\beta\} \vdash \neg\beta$ and by Theorem 3.7 $B \vdash \beta \to \neg\beta$. Finally, using axiom 9 we have $\neg\beta \in B$. Conversely if $\neg\beta \in B$ then, by consistency $\beta \notin B$.

($\to$) $B \vDash_M \beta_1 \to \beta_2$ iff, by induction hypothesis, for every full extension B' of B we have that $\beta_1 \notin B'$ or $\beta_2 \in B'$. First assume that $B \cup \{\beta_1\}$ is inconsistent, then $B \cup \{\beta_1\} \vdash \beta_2$ by Proposition 3.8. Now assume that $B \cup \{\beta_1\}$ is consistent and that $B \cup \{\beta_1\} \nvdash \beta_2$, then by Lemma 3.19 there is an full extension of $B \cup \{\beta_1\}$ that do not derive β_2 which is a contradiction. It follows that $B \cup \{\beta_1\} \vdash \beta_2$ and by Theorem 3.7 $\beta_1 \to \beta_2 \in B$.

Conversely, if for all full extension B' of B it holds that $\beta_1 \to \beta_2 \in B'$ then if $\beta_1 \in B'$ by modus ponens $\beta_2 \in B'$. Hence $B' \nvDash \beta_1$ or $B' \vDash \beta_2$ and we conclude that $B \vDash \beta_1 \to \beta_2$.

To conclude the proof, just notice that $B \nvDash_M \alpha$. $\qquad\square$

3.2.3 Properties

By Theorems 3.18 and 3.20, the consequence operator C_{Int} induced either by the syntactic consequence relation $\vdash_{Int}$ or by the semantic consequence relation $\vDash_{Int}$ are the same, i.e.,

$$C_{Int}(A) = \{\alpha \in \mathscr{L} : A \vdash_{Int} \alpha\} = \{\alpha \in \mathscr{L} : A \vDash_{Int} \alpha\}$$

Int is distributive, but it is not decomposable.

Proposition 3.21 *Int is distributive.*

Proof First notice that if $\beta \in Cn(\gamma_1) \cap Cn(\gamma_2)$ then by deduction $\gamma_1 \to \beta, \gamma_2 \to \beta \in Cn(\emptyset)$. Using Axiom 10 we have that $\beta \in Cn(\gamma_1 \vee \gamma_2)$. It follows that $Cn(\gamma_1) \cap Cn(\gamma_2) \subseteq Cn(\gamma_1 \vee \gamma_2)$.

Using the semantic definition it is trivial to verify that $Cn(\{\beta_1, \ldots, \beta_n\}) = Cn(\beta_1 \wedge \cdots \wedge \beta_n)$. Since Int is compact, the rest of the proof follows exactly as in Lemma 2.13. $\qquad\square$

Proposition 3.22 *Int is not decomposable.*

Proof Let $K = \{\neg p\}$ and $A = \{\neg p \vee p\}$. We will show that: 1) $C_{Int}(\emptyset) \subset C_{Int}(A) \subset C_{Int}(K)$ and 2) There is no X such that $C_{Int}(X) \subset C_{Int}(K)$ and $C_{Int}(X \cup A) = C_{Int}(K)$

1. Using axiom 6 we have that $p \vee \neg p \in C_{Int}(\neg p)$. Consider the following Kripke model $M = \langle W, \leq, v \rangle$:

$$W = \{w\}$$
$$\leq = \{(w, w)\}$$
$$v(p) = \{w\}$$

It is easy to verify that $\vDash_M p \vee \neg p$, but $\nvDash_M \neg p$. It follows that $C_{Int}(p \vee \neg p) \neq C_{Int}(\neg p)$.

Furthermore, Example 3.16 shows that $p \vee \neg p \notin C_{Int}(\emptyset)$.

2. We will show that for any $X \subseteq \mathscr{L}$ such that $C_{Int}(X) \subset C_{Int}(\neg p)$ there is a model M such that $\vDash_M X \cup \{p \vee \neg p\}$, but $\nvDash_M \neg p$.

   Take X such that $C_{Int}(X) \subset C_{Int}(\neg p)$. Since $\neg p \notin C_{Int}(X)$ there is a model $M' = \langle W', \leq', v' \rangle$ such that $\vDash_{M'} X$ and there is a $w \in W'$ with $w_0 \leq w$ and $w \in v'(p)$. Now, consider the following model $M = \langle W, \leq, v \rangle$

$$W = \{w \in W' : w \in v'(p)\}$$
$$\leq = \{(w_1, w_2) \in W \times W : w_1 \leq' w_2\}$$
$$v(p) = v'(p) \cap W$$

First notice that $\leq$ is indeed a partial order. Then notice that $\vDash_M X$, $\vDash_M p$. Of course, we also have that $\vDash_M p \vee \neg p$. It follows that $X \cup \{p \vee \neg p\} \nvDash_M \neg p$, hence $X \cup \{p \vee \neg p\} \nvDash \neg p$.

In Intuitionistic logic the connective $\neg$ is not classical and the logic is not complemented. Hence, Int fails to satisfy AGM-assumptions. $\square$

3.3 Horn Logic (Horn)

The problem of verifying if a sentence α is a consequence of a set of sentences A in Classical Propositional Logic is hard (CoNP-complete). One way to avoid such complexity is to restrict the language $\mathscr{L}_{CPL}$. *Horn logic* is a fragment of CPL whose inferential problem can be solved in polynomial time.

Of course when the language is restricted it became less expressive. Although, Horn logic is widely used in computer science, because of its low computational complexity. Several applications in Artificial Intelligence uses Horn logic. Programing language PROLOG[1], for example, is based on Horn logic.

Consider an enumerable set of propositional variables $\mathbb{P}$. The language $\mathscr{L}_{Horn}$ of Horn logic is defined using BNF as follows:

$$\begin{aligned} \text{head} &:= p & p \in \mathbb{P} \\ \text{body} &:= \neg p \mid (\text{body} \vee \text{body}) & p \in \mathbb{P} \\ \text{clause} &:= \text{head} \mid \text{body} \mid (\text{head} \vee \text{body}) \\ \alpha &:= \text{clause} \mid (\alpha \wedge \alpha) \end{aligned}$$

---

[1] http://www.swi-prolog.org

A sentence $\alpha \in \mathcal{L}_{\text{Horn}}$ is a conjunction of Horn clauses. A Horn clause has a body (a disjunction of propositional variables) and may have one head (a negated propositional variable). Of course we have that $\mathcal{L}_{\text{Horn}} \subset \mathcal{L}_{\text{CPL}}$.

Example 3.23

$$p \in \mathcal{L}_{\text{Horn}}$$
$$\neg p \in \mathcal{L}_{\text{Horn}}$$
$$p_1 \vee \neg p_2 \in \mathcal{L}_{\text{Horn}}$$
$$(p_1 \vee \neg p_2) \wedge (p_2 \vee \neg p_3) \in \mathcal{L}_{\text{Horn}}$$
$$p_1 \vee p_2 \notin \mathcal{L}_{\text{Horn}}$$

The consequence operator C_{Horn} is simply the restriction of C_{CPL} to the language $\mathcal{L}_{\text{Horn}}$, i.e.,

$$C_{\text{Horn}}(A) = \{\alpha \in \mathcal{L}_{\text{Horn}} : \alpha \in C_{\text{CPL}}(A)\}$$

A *query* in Horn logic is a sentence α with no head. The most important inferential problem in Horn logic is to decide whether or not a query α follows from a set $A \subseteq \mathcal{L}_{\text{Horn}}$. Let $\alpha = \neg p_1 \vee \cdots \vee \neg p_n$ be a query and consider the set $A' = A \cup \{p_1, \ldots, p_n\}$. This problem can be reduced to the problem of deciding whether there is a truth-function v such that $v(A') = 1$, i.e., deciding whether A' is satisfiable (Horn-SAT). Horn-SAT can be solved in polynomial time.

3.3.1 Properties

Horn logic is compact, but it is neither decomposable, distributive nor complemented.

Proposition 3.24 *Horn logic is compact*

Proof Let $A \subseteq \mathcal{L}_{\text{Horn}}$. Since $\mathcal{L}_{\text{Horn}} \subseteq \mathcal{L}_{\text{CPL}} \subseteq \mathcal{L}_{\text{CPL}}$. By compactness of CPL we have that there is a finite $A' \subseteq A$ such that $\alpha \in C_{\text{CPL}}(A')$. Of course $A' \subseteq \mathcal{L}_{\text{Horn}}$ and, hence, $\alpha \in C_{\text{Horn}}(A')$. $\qquad\square$

Proposition 3.25 *Horn logic is not decomposable.*

Proof Let $Y = Cn(\{\beta \in \mathcal{L} : Cn(\beta) \subset Cn(p)\})$. For every $p_i \in \mathbb{P}$, we have $\neg p_i \vee p \in C_{\text{Horn}}(Y)$. It follows that $C_{\text{Horn}}(Y) \neq C_{\text{Horn}}(\emptyset)$. Consider a sentence β such that $C_{\text{Horn}}(\beta) \subset C_{\text{Horn}}(p)$. Every clause in this sentence has at least one $\neg p_i$ for $p_i \neq p$, otherwise either $\beta \notin C_{\text{Horn}}(p)$ or $C_{\text{Horn}}(\beta) = C_{\text{Horn}}(p)$. Consider a truth-function v such that $v(p) = 0$ for every $p \in \mathbb{P}$. Since every clause in β has at least one $\neg p_i$ then $\models_v \beta$ and, hence, $\models_v Y$. Since $\nvDash_v p$, we have that $C_{\text{Horn}}(Y) \neq C_{\text{Horn}}(p)$. Of course $C_{\text{Horn}}(Y) \subseteq C_{\text{Horn}}(p)$. Hence, $C_{\text{Horn}}(Y) \subset C_{\text{Horn}}(p)$. By Lemma 2.9, $\langle \mathcal{L}, Cn \rangle$ is not decomposable. $\qquad\square$

Proposition 3.26 *Horn logic is not distributive.*

Proof

$$p \in C_{\text{Horn}}(\{p \vee \neg q, p\}) \cap C_{Hron}(\{p \vee \neg q, q\})$$
$$p \notin C_{\text{Horn}}(\{p \vee \neg q\} \cup (C_{\text{Horn}}(p) \cap C_{\text{Horn}}(q)))$$
$$= C_{\text{Horn}}(\{p \vee \neg q\} \cup C_{\text{Horn}}(\emptyset))$$
$$= C_{\text{Horn}}(\{p \vee \neg q\}) \qquad \qquad \Box$$

Proposition 3.27 *Horn logic is not complemented.*

Proof The sentence $\neg p \wedge \neg q \in \mathscr{L}_{\text{Horn}}$ has no complement. There is no A such that $C_{\text{Horn}}(A) \neq C_{\text{Horn}}(\emptyset)$ and $C_{\text{Horn}}(\neg p \wedge \neg q) \cap C_{\text{Horn}}(A) = C_{\text{Horn}}(\emptyset)$. $\qquad \Box$

3.4 Description Logics

Description Logics (DLs) are formalisms used to represent relations between concepts and between concepts and instances in a domain. Differently from their predecessors (frame systems [Min81] and semantic networks [Qui67]), DLs have well-defined semantics based on first-order logic. In fact, DLs are fragments of first-order logic with the advantage of having decidable inference problems.

When modeling a domain it is desirable, on the one hand, an expressive formalism capable of representing a vast amount of different problems. On the other hand, it is important to guaranty that the inference problems can be solved efficiently. Both desiderata are, sometimes, incompatible (see [LB87]). Finding a good balance between expressivity and computational complexity is one of the main challenges in knowledge representation.

The research in DLs explored the complexity and expressivity of a wide range of logics and culminated with the adoption of OWL as the standard language for representing ontologies in the Web. Both versions of OWL were deeply inspired in DLs.

Ontology following Grubber [Gru93] is a formal "specification of a conceptualization". Ontologies in computer science gain attention with the advent of semantic Web. *Semantic Web* has been pointed as the evolution of the Web where available information would be comprehensive by humans as well as artificial agents.

The role of ontologies in the Web would be to provide common vocabulary between web-agents. In semantic Web, ontologies would be responsible to declare the terms used by each agent and to make explicit how this terms are semantically related. In a few words, ontologies would enable a richer communication between artificial agents.

DLs are a *family* of logics and each DL has its own language, semantics, etc. In what follows the language and the semantic consequence relation for a specific DL will be presented, $\mathcal{ALC}$. After that extensions of $\mathcal{ALC}$ will be presented. The section finishes presenting OWL, the standard language to represent ontologies on the Web.

3.4.1 Language

Before defining the language of $\mathcal{ALC}$ we need to present its *description language*. The description language consists of the concepts of the logic.

The signature of a description language is a tuple $\langle N_C, N_R, N_I \rangle$ such that N_C, N_R and N_I are infinite enumerable sets of atomic concepts, atomic roles, and individuals, respectively. The following is a presentation of the description language of $\mathcal{ALC}$ in BNF ($A \in N_C$ and $R \in N_R$):

$$C := A \mid \top \mid \bot \mid (\neg C) \mid (C \sqcap C) \mid (C \sqcup C) \mid (\forall R.C) \mid (\exists R.C)$$

Symbols $\neg$, $\sqcap$, $\sqcup$, $\forall$ and $\exists$ are called *constructors*. A description language is characterized by the constructors it admits. $\mathcal{ALC}$, for example, admits negation ($\neg$), conjunction ($\sqcap$), disjunction ($\sqcup$), value restriction ($\forall$) and qualified existential restriction ($\exists$).

Example 3.28 Let Man, Mortal be atomic concepts and is Disciple Of a role. The following are concepts in $\mathcal{ALC}$:

Man: the set of men.
Man $\sqcap$ Mortal: the set of mortal men.
Man $\sqcap \neg$Mortal: the set of immortal men.
Man $\sqcap \exists isDiscipleOf.$Man: the set of disciples.

The language of $\mathcal{ALC}$ is composed of two parts: the TBox and the ABox. TBox contains sentences that represent conceptual knowledge while the ABox contains sentences that represent assertional knowledge.

Sentences in $\mathcal{L}_{\mathcal{ALC}}^{TBox}$ (the TBox of $\mathcal{ALC}$) has the following form:

$$\alpha := C \sqsubseteq C$$

Sentence in $\mathcal{L}_{\mathcal{ALC}}^{ABox}$ (the ABox of $\mathcal{ALC}$) has the following form where C is a concept, $R \in N_R$ and $a, b \in N_I$:

$$\alpha := C(a) \mid R(a, b) \mid a = b \mid a \neq b$$

Example 3.29 Let Man, Mortal $\in N_C$ and let SOCRATES, PLATO be instances i.e. elements of N_I. The following are sentences in $\mathcal{ALC}$ language $\mathcal{L}_{\mathcal{ALC}} = \mathcal{L}_{\mathcal{ALC}}^{TBox} \cup \mathcal{L}_{\mathcal{ALC}}^{ABox}$:

Man $\sqsubseteq$ Mortal
Man(SOCRATES)
(Man $\sqcap \neg$Mortal)$(SOCRATES)$
isDiscipleOf(SOCRATES, PLATO)

3.4.2 Semantics

An *interpretation* is a tuple $\mathcal{I} = \langle \Delta^{\mathcal{I}}, \cdot^{\mathcal{I}} \rangle$ such that $\Delta^{\mathcal{I}}$ is a non-empty set called *domain* and $\cdot^{\mathcal{I}}$ is a function called *interpretation function*. The interpretation function associates atomic concepts to subsets of the domain, atomic roles to binary relations on the domain, and individuals with elements of the domain. Formally:

- If $A \in N_C$ then $A^{\mathcal{I}} \subseteq \Delta^{\mathcal{I}}$.
- If $R \in N_R$ then $R^{\mathcal{I}} \subseteq \Delta^{\mathcal{I}} \times \Delta^{\mathcal{I}}$.
- If $a \in N_I$ then $a^{\mathcal{I}} \in \Delta^{\mathcal{I}}$.

The interpretation is extended for complex concepts as follows:

$$(\top)^{\mathcal{I}} = \Delta^{\mathcal{I}}$$
$$(\bot)^{\mathcal{I}} = \emptyset$$
$$(\neg C)^{\mathcal{I}} = \Delta^{\mathcal{I}} \setminus C^{\mathcal{I}}$$
$$(C \sqcap D)^{\mathcal{I}} = C^{\mathcal{I}} \cap D^{\mathcal{I}}$$
$$(C \sqcup D)^{\mathcal{I}} = C^{\mathcal{I}} \cup D^{\mathcal{I}}$$
$$(\forall R.C)^{\mathcal{I}} = \{a : (a, b) \in R^{\mathcal{I}} \rightarrow b \in C^{\mathcal{I}}\}$$
$$(\exists R.C)^{\mathcal{I}} = \{a : \exists (a, b) \in R^{\mathcal{I}} \wedge b \in C^{\mathcal{I}}\}$$

Example 3.30 Consider the following interpretation $\mathcal{I} = \langle \Delta^{\mathcal{I}}, \cdot^{\mathcal{I}} \rangle$:

$$\Delta^{\mathcal{I}} = \{s, p, z\}$$
$$\text{Man} = \{s, p\}$$
$$\text{Mortal} = \{p\}$$

This interpretation is extended to complex concepts as follows:

$$\begin{aligned}
(\text{Man} \sqcap \neg\text{Mortal})^{\mathcal{I}} &= \text{Man}^{\mathcal{I}} \cap (\neg\text{Mortal})^{\mathcal{I}} \\
&= \text{Man}^{\mathcal{I}} \cap (\Delta^{\mathcal{I}} \setminus \text{Mortal})^{\mathcal{I}} \\
&= \{s, p\} \cap (\{s, p, z\} \setminus \{p\}) \\
&= \{s\}
\end{aligned}$$

An interpretation $\mathcal{I}$ satisfies (or is a model for) a sentence $\alpha \in \mathcal{L}_{\mathcal{ALC}}$ (written $\models_{\mathcal{I}} \alpha$) iff it holds that:

$$\begin{aligned}
&\models_{\mathcal{I}} C \sqsubseteq D \text{ iff } C^{\mathcal{I}} \subseteq D^{\mathcal{I}} \\
&\models_{\mathcal{I}} C(a) \quad \text{ iff } a^{\mathcal{I}} \in C^{\mathcal{I}} \\
&\models_{\mathcal{I}} R(a, b) \text{ iff } (a^{\mathcal{I}}, b^{\mathcal{I}}) \in R^{\mathcal{I}} \\
&\models_{\mathcal{I}} a = b \quad \text{ iff } a^{\mathcal{I}} = b^{\mathcal{I}} \\
&\models_{\mathcal{I}} a \neq b \quad \text{ iff } a^{\mathcal{I}} \neq b^{\mathcal{I}}
\end{aligned}$$

As usual, an interpretation satisfies a set A (in symbols $\models_{\mathcal{I}} A$) iff A satisfies each element of the set, i.e., $\models_{\mathcal{I}} A$ iff $\models_{\mathcal{I}} \beta$ for every $\beta \in A$. A a sentence α is a *semantic consequence* of set of sentences A (written A iff every model of A is a model of α, i.e., for every $\mathcal{I}$ if $\models_{\mathcal{I}} A$ then $\models_{\mathcal{I}} \alpha$.

Example 3.31 Let us show that

$$\{\text{Man} \sqsubseteq \text{Mortal}, \text{Man}(\text{SOCRATES})\} \models \text{Mortal}(\text{SOCRATES})$$

Proof For every $\mathcal{I}$ if $\models_{\mathcal{I}} \text{Man}(\text{SOCRATES})$ and $\models_{\mathcal{I}} \text{Man} \sqsubseteq \text{Mortal}$ then we have:

$$\text{SOCRATES}^{\mathcal{I}} \in \text{Man}^{\mathcal{I}}$$
$$\text{Man}^{\mathcal{I}} \quad\quad \subseteq \text{Mortal}^{\mathcal{I}}$$

Hence, $\text{SOCRATES}^{\mathcal{I}} \in \text{Mortal}^{\mathcal{I}}$.

The consequence operator is defined as usual $Cn(A) = \{\alpha \in \mathscr{L}_{\mathcal{ALC}} : A \models \alpha\}$ and it is Tarskian. Remember that two sets of sentences A and B are equivalent iff $Cn(A) = Cn(B)$, i.e., A and B are equivalents iff for all interpretations $\mathcal{I}$ we have that $\models_{\mathcal{I}} A$ iff $\models_{\mathcal{I}} B$.

3.4.3 Extensions of $\mathcal{ALC}$

Several extensions for $\mathcal{ALC}$ has been proposed in the literature [BCM+03]. We will show some of these extensions, namely, transitive roles, inverse roles, role hierarchy, nominals, and numeric restrictions.

A DL that admits *transitive roles*, besides the constructors for $\mathcal{ALC}$, is called $\mathcal{S}$. In $\mathcal{S}$ transitive roles form a set $N_{R_+} \subseteq N_R$, i.e., for $R \in N_{R_+}$ we have that for every $\mathcal{I}$:

$$\text{if } (x, y) \in R^{\mathcal{I}} \text{ and } (y, z) \in R^{\mathcal{I}} \text{ then } (x, z) \in R^{\mathcal{I}}$$

The names of $\mathcal{ALC}$ extensions are build adding letters to $\mathcal{ALC}$ or to $\mathcal{S}$. For example $\mathcal{SI}$ is the DL obtained extending $\mathcal{ALC}$ with transitive roles and inverse roles. Consider the set $N_R \cup \{R^- : R \in N_R\}$. For every $\mathcal{I}$ and every $R \in N_R \cup \{R^- : R \in N_R\}$:

$$\text{if } (x, y) \in R^{\mathcal{I}} \text{ then } (y, x) \in (R^-)^{\mathcal{I}}$$

If the logic admits transitive roles then $R \in N_{R_+}$ implies that $R^- \in N_{R_+}$. Furthermore, in order to avoid using symbols like R^-, we will use the following abbreviation:

$$Inv(R) = \begin{cases} R^- & \text{if } R \in N_R \\ S & \text{if } R = S^- \end{cases}$$

DLs that admits role hierarchy are represented with the letter $\mathcal{H}$ in their name. Such logics have, besides the TBox and the ABox, an RBox that contains sentences like:

$$\alpha := R \sqsubseteq S$$

Let $\mathcal{R}$ be an RBox, define the relation $\tilde{\sqsubseteq}$ as the reflexive transitive closure of $\sqsubseteq$ on $\mathcal{R} \cup \{Inv(R) \sqsubseteq Inv(S) : R \sqsubseteq S \in \mathcal{R}\}$ and let $\mathcal{R}^+$ be the set of axiom of the type $R \sqsubseteq S$ such that $R\tilde{\sqsubseteq}S$. The abbreviation $R \equiv S$ will be used to indicate that $R\tilde{\sqsubseteq}S$ and $S\tilde{\sqsubseteq}R$.

If $R \in N_{R_+}$ and $R \equiv S$ then S is transitive. The symbol $Trans(R)$ will be used to denote that R is transitive, i.e., $Trans(R)$ iff $R \equiv S$ for some $S \in N_{R_+}$ or $Inv(S) \in N_{R_+}$. Non transitive roles are called *simple*.

An interpretation $\mathcal{I}$ satisfies the RBox $\mathcal{R}$ (written $\vDash_{\mathcal{I}} \mathcal{R}$) iff for every $R \sqsubseteq S \in \mathcal{R}$:

$$R^{\mathcal{I}} \subseteq S^{\mathcal{I}}$$

DLs that admits *nominals* are represented with the letter $\mathcal{O}$. In these DLs it is possible to define concepts enumerating its individuals ($C = \{a_1, \ldots, a_n\}$ for $a_1, \ldots, a_n \in N_I$). The interpretation of these concepts are as expected:

$$\{a_1, \ldots, a_n\}^{\mathcal{I}} = \{a_1^{\mathcal{I}}, \ldots, a_n^{\mathcal{I}}\}$$

In DLs that admit nominals, any ABox sentence has an equivalent sentence in the TBox:

$$
\begin{array}{ll}
C(a) & \text{iff } \{a\} \sqsubseteq C \\
R(a, b) & \text{iff } \{a\} \sqsubseteq \exists R.\{b\} \\
a = b & \text{iff } \{a\} \sqsubseteq \{b\} \\
a \neq b & \text{iff } \{a\} \sqsubseteq \neg\{b\}
\end{array}
$$

A DL admits *qualified numeric restriction* ($\mathcal{Q}$) iff it admits the constructions $\leq_n R.C$ and $\geq_n R.C$ for any simple (non-transitive) role R, any concept C, and any positive integer n. The concept $\leq_n R.C$ represents the set whose elements are related via R with at most n elements of C. Formally, using the symbol # to represent the cardinality of a set:

$$(\leq R.C)^{\mathcal{I}} = \{b : \#\{(b, a) \in R^{\mathcal{I}} \wedge a \in C^{\mathcal{I}}\} \leq n\}\}$$
$$(\geq R.C)^{\mathcal{I}} = \{b : \#\{(b, a) \in R^{\mathcal{I}} \wedge a \in C^{\mathcal{I}}\} \geq n\}\}$$

The letter $\mathcal{N}$ is used for logics that admit only *simple numeric restrictions*, i.e., concepts of the form $\leq_n R.\top$ and $\geq_n R.\top$. If, besides that, n is restricted to 1 then we say that the logic admits *functional roles* ($\mathcal{F}$).

In many situations we need to reason not only about abstract classes such as Mortal or Man, but also with data types like string and integer. Data types have its own domain $\Delta_D^{\mathcal{I}}$ disjoint from $\Delta^{\mathcal{I}}$. A *concrete role* is a binary relation in $\Delta^{\mathcal{I}} \times \Delta_D^{\mathcal{I}}$. Concrete roles can be used in numeric restrictions, existential restrictions, and value restrictions. A DL that admits concrete roles is designed by the letter D between parenthesis.

Table 3.1 sums up the semantics of the DLs presented in this section.

Table 3.1 $\mathcal{ALC}$ extensions

Data types (**D**)	D	$D^{\mathcal{I}} \subseteq \Delta_{\mathrm{D}}^{\mathcal{I}}$
Concrete roles	U	$U^{\mathcal{I}} \subseteq \Delta^{\mathcal{I}} \times \Delta_{\mathrm{D}}^{\mathcal{I}}$
Data value	v	$v^{\mathcal{I}} \in \Delta_{\mathrm{D}}^{\mathcal{I}}$
Existential restriction	$\exists U.D$	$\{x : \exists (x, y) \in U^{\mathcal{I}} \wedge y \in D^{\mathcal{I}}\}$
Value restriction	$\forall U.D$	$\{x : (x, y) \in U^{\mathcal{I}} \rightarrow y \in D^{\mathcal{I}}\}$
Functional roles ($\mathcal{F}$)	$\leq_1 R$	$\{x : \#\{(x, y) \in R^{\mathcal{I}}\} \leq 1\}$
	$\leq_1 U$	$\{x : \#\{(x, y) \in U^{\mathcal{I}}\} \leq 1\}$
	$\geq_1 R$	$\{x : \#\{(x, y) \in R^{\mathcal{I}}\} \geq 1\}$
	$\geq_1 U$	$\{x : \#\{(x, y) \in U^{\mathcal{I}}\} \geq 1\}$
Simple numeric restr. ($\mathcal{N}$)	$\leq_n R$	$\{x : \#\{(x, y) \in R^{\mathcal{I}}\} \leq n\}$
	$\leq_n U$	$\{x : \#\{(x, y) \in U^{\mathcal{I}}\} \leq n\}$
	$\geq_n R$	$\{x : \#\{(x, y) \in R^{\mathcal{I}}\} \geq n\}$
	$\geq_n U$	$\{x : \#\{(x, y) \in U^{\mathcal{I}}\} \geq n\}$
Qualified numeric restr. ($\mathcal{Q}$)	$\leq_n R.C$	$\{x : \#\{(x, y) \in R^{\mathcal{I}} \wedge y \in C\} \leq n\}$
	$\leq_n U.D$	$\{x : \#\{(x, y) \in U^{\mathcal{I}} \wedge y \in D\} \leq n\}$
	$\geq_n R.C$	$\{x : \#\{(x, y) \in R^{\mathcal{I}} \wedge y \in C\} \geq n\}$
	$\geq_n U.D$	$\{x : \#\{(x, y) \in U^{\mathcal{I}} \wedge y \in D\} \geq n\}$
Nominals ($\mathcal{O}$)	$\{a_1, \ldots, a_n\}$	$\{d : d = a_i^{\mathcal{I}}\}$
Inverse roles ($\mathcal{I}$)	R^-	$\{(x, y) : (y, x) \in R^{\mathcal{I}}\}$
Transitive roles ($\mathcal{S}$)	$R \in R_+$	$\{R^{\mathcal{I}} = (R^{\mathcal{I}})^+\}$
Role hierarchy ($\mathcal{H}$)	$R \sqsubseteq S$	$R^{\mathcal{I}} \subseteq S^{\mathcal{I}}$

3.4.4 OWL

In 2004 W3C[2] officially announced the adoption of OWL as the standard language for representing ontologies on the Web.

This announcement was the climax of several years of studies. During these years one of the goals of the DL researchers was to study very expressive decidable fragments of first-order logic. Furthermore, in the end of the 1990s and beginning of the twenty-first century several inference engines for expressive DLs were developed: RACER [HM01] and FaCT [Hor98] are two good examples. Those inference engines based in tableaux methods had a surprisingly good performance for such expressive formalisms.

For these reasons, the development of OWL was deeply influenced by Description Logics field. The first version of OWL came in three flavors with decreasing expressive powers and computational complexity: OWL-full, OWL-DL, and OWL-lite. Although completely compliant with RDFS, OWL-full is undecidable. OWL-DL is a syntactic variant of the Description Logic $\mathcal{SHOIN}$(D), i.e., the DL that extends $\mathcal{ALC}$ with transitive roles ($\mathcal{S}$), role hierarchy ($\mathcal{H}$), nominals ($\mathcal{O}$), inverse roles ($\mathcal{I}$), simple numeric restrictions ($\mathcal{N}$) and datatypes (D). Besides its enormous computational complexity, modern inference engines such as Hermit

---

[2] http://www.w3c.org

[MSH07], FaCT++ [TH06], and RACER [HM01] are capable of solving inference problems in $\mathcal{SHOIN}$(D) surprisingly fast in most cases. Finally, OWL-lite is a syntactic variant of the logic $\mathcal{SHIF}$(D) which is a bit less expressive then $\mathcal{SHOIN}$(D). The logic $\mathcal{SHIF}$(D) extends $\mathcal{ALC}$ with role hierarchy, functionals, inverse and transitive roles, and datatypes. It turns out, however, that inference problems in $\mathcal{SHIF}$(D) still had high complexity (ExpTime complete).

In 2009, a new version of OWL was released [OWL09]. This time OWL DL was a syntactic variation of a DL called $\mathcal{SROIQ}$. $\mathcal{SROIQ}$ extends $\mathcal{SHOIN}$(D) with the following new functionalities:

- *disjoint roles:* a new type of RBox sentence to express that two roles are disjoint.
- *reflexive and irreflexive roles:* new types of RBox sentences to express that a roles is (ir)reflexive.
- *negated role assertions:* a new type of ABox sentence to express that certain individuals are not related via certain relation.
- *complex role hierarchy:* new types of RBox sentences ($R \circ S \sqsubseteq R$ and $S \circ R \sqsubseteq R$ that guarantees that R contains the composition of R and S and the composition of S and R, respectively.
- *universal role:* a new constant U such that $U^{\mathcal{I}} = \Delta^{\mathcal{I}} \times \Delta^{\mathcal{I}}$ for every interpretation $\mathcal{I}$.
- *Self:* a concept that allows self-reference, e.g., the following is a definition of the concept narcissist as someone who likes herself $Narcissist \equiv likes.Self$.

The second version of OWL also defines tractable fragments, i.e., sublanguages whose inference problems can be solved polynomially. This fragments are called OWL-profiles [MGH+08]. Each profile was defined for a different purpose:

1. *OWL-EL* is appropriate to represent ontologies with many simple concepts. Besides its low expressivity, OWL-EL is enough to model satisfactorily important medical ontologies such as SNOMED [SR04] and large part of GALEN [RRP96]. The theoretical framework of OWL-EL is the $\mathcal{EL}$ family of DLs [BBL08].
2. *OWL-QL* is appropriate for ontologies with many instances such that the main use is to answer queries. Besides the low expressivity, in OWL-QL it is possible to model the main characteristics of UML and ER. The theoretical framework of OWL-QL is the $\mathcal{DL}$-lite family [ACKZ09].
3. *OWL-RL* is appropriate for applications that need high computational complexity. The efficiency of OWL-RL comes from the fact that it can be implemented using rules.

3.4.5 Properties

Since DLs are fragments of first-order logic, using the same argument from Proposition 3.24, we have that DLs are compact. Other properties hold for certain DLs. To facilitate our exposure we will split the DLs in two categories: regulars and irregulars.

A DL $\langle \mathscr{L}, Cn \rangle$ is *regular* iff every sentence $\alpha \in \mathscr{L}$ is equivalent to a sentence in the form:

$$\top \sqsubseteq B$$

Proposition 3.32 *If a DL $\langle \mathscr{L}, Cn \rangle$ admits only TBox sentences and admits negation and conjunction then $\langle \mathscr{L}, Cn \rangle$ is regular.*

Proof Any sentence $C \sqsubseteq D$ is equivalent to $\top \sqsubseteq \neg C \sqcup D$.   □

Proposition 3.33 *If a DL $\langle \mathscr{L}, Cn \rangle$ admits TBox and ABox sentences and admits negation, conjunction, existential restriction, and nominals then $\langle \mathscr{L}, Cn \rangle$ is regular.*

Proof As already argued, if $\langle \mathscr{L}, Cn \rangle$ admits nominals then every sentence in the ABox is equivalent to some sentence in the TBox.   □

Using Proposition 3.33 it is easy to verify which DLs are regular and which are not. The following DLs are *regular*:

- Any logic between $\mathcal{ALC}$ and $\mathcal{ALCIQ}$ with empty ABox.
- Any logic between $\mathcal{S}$ and $\mathcal{SIQ}$ with empty ABox.
- Any logic between $\mathcal{ALCO}$ and $\mathcal{ALCOIQ}$.
- Any logic between $\mathcal{SO}$ and $\mathcal{SOIQ}$.

The following DLs are *irregular*:

- Any logic between $\mathcal{ALC}$ and $\mathcal{ALCIQ}$.
- Any logic between $\mathcal{S}$ and $\mathcal{SIQ}$.
- Any logic between $\mathcal{ALCH}$ and $\mathcal{ALCOIOH}$.
- Any logic between $\mathcal{SH}$ and $\mathcal{SHOIN}$(D) including $\mathcal{SHIF}$(D).
- The logic $\mathcal{SROIQ}$.

Lemma 3.34 *If $\langle \mathscr{L}, Cn \rangle$ is compact and for every finite $A, B, C \in 2^{\mathscr{L}}$ it holds that*

$$Cn(A \cup B) \cap Cn(A \cup C) \subseteq Cn(A \cup (Cn(B) \cap Cn(C)))$$

then $\langle \mathscr{L}, Cn \rangle$ is distributive.

Proof Let $\beta \in Cn(A \cup B) \cap Cn(A \cup C)$ then there are finite sets A_1 and A_2 such that $A_1 \subseteq A \cup B$, $A_2 \subseteq A \cup C$ and $\beta \in Cn(A_1) \cap Cn(A_2)$. Let $B' = B \cap A_1$, $C' = C \cap A_2$ and $A' = A \cap (A_1 \cup A_2) = (A \cap A_1) \cup (A \cap A_2)$. It follows that $B' \subseteq B$, $C' \subseteq C$ and $A' \subseteq A$. With a simple set manipulation we can show that $Cn(A_1) \cap Cn(A_2) \subseteq Cn(A' \cup B') \cap Cn(A' \cup C')$. By hypothesis $Cn(A' \cup B') \cap Cn(A' \cup C') \subseteq Cn(A' \cup (Cn(B' \cap Cn(C'))))$ and we conclude that $\beta \in Cn(A \cup (Cn(B \cap Cn(C))))$.   □

Proposition 3.35 *Any regular DL that admits* $\sqcup$ *and* $\sqcap$ *is distributive.*[3]

Proof By Lemma 3.34 we need to prove just the finite case and since the logic is regular we can assume without loss of generality that $\mathcal{B} = \{\top \sqsubseteq B_1, \ldots, \top \sqsubseteq B_m\}$ and $\mathcal{C} = \{\top \sqsubseteq C_1, \ldots, \top \sqsubseteq C_n\}$.

Let $B = B_1 \sqcap \cdots \sqcap B_m$ and $C = C_1 \sqcap \ldots C_n$. Of course $Cn(\mathcal{B}) = Cn(\top \sqsubseteq B)$ and $Cn(\mathcal{C}) = Cn(\top \sqsubseteq C)$. Let $\beta \in Cn(\mathcal{A} \cup \mathcal{B}) \cap Cn(\mathcal{A} \cup \mathcal{C}) = Cn(\mathcal{A} \cup \{\top \sqsubseteq B\}) \cap Cn(\mathcal{A} \cup \{\top \sqsubseteq C\})$.

Now notice that the second line below follows from the first.

$$\forall \mathcal{I}(\vDash_{\mathcal{I}} \mathcal{A} \text{ and } (\vDash_{\mathcal{I}} \top \sqsubseteq B \text{ or } \vdash_{\mathcal{I}} \top \sqsubseteq C) \Rightarrow \vDash_{\mathcal{I}} \alpha)$$
$$\forall \mathcal{I}(\vDash_{\mathcal{I}} \mathcal{A} \text{ and } \vDash_{\mathcal{I}} \top \sqsubseteq B \sqcap C \qquad\qquad \Rightarrow \vDash_{\mathcal{I}} \alpha)$$

It follows that $Cn(\mathcal{A} \cup \mathcal{B}) \cap Cn(\mathcal{A} \cup \mathcal{C}) \subseteq Cn(\mathcal{A} \cup \{\top \sqsubseteq B \sqcap C\}$.

Notice also that the second and third lines below follows from the first.

$$\forall \mathcal{I}(\vDash_{\mathcal{I}} \top \sqsubseteq B \sqcup C \Rightarrow \vDash_{\mathcal{I}} \alpha)$$
$$\forall \mathcal{I}(\vDash_{\mathcal{I}} \top \sqsubseteq B \qquad \Rightarrow \vDash_{\mathcal{I}} \alpha)$$
$$\forall \mathcal{I}(\vDash_{\mathcal{I}} \top \sqsubseteq C \qquad \Rightarrow \vDash_{\mathcal{I}} \alpha)$$

Hence, $Cn(\mathcal{A} \cup \{\top \sqsubseteq B \sqcap C\}) \subseteq Cn(\mathcal{A} \cup (Cn(\mathcal{B}) \cap Cn(\mathcal{C})))$. We conclude that $Cn(\mathcal{A} \cup \mathcal{B}) \cap Cn(\mathcal{A} \cup \mathcal{B}) \subseteq Cn(\mathcal{A} \cup (Cn(\mathcal{B}) \cap Cn(\mathcal{C})))$ □

Many irregular DLs, though, are not distributive. Consider two examples:

Example 3.36 Let $X = \{a = b\}$, $Y = \{C(a)\}$ and $Z = \{C(b)\}$. Since $Cn(Y) \cap Cn(Z) = Cn(\emptyset))$, it follows that $C(a) \notin Cn(X \cup (Cn(Y) \cap Cn(Z)))$. However, $C(a) \in Cn(X \cup Y) \cap Cn(X \cup Z)$. We conclude that $\mathcal{ALC}$ is not distributive.

Example 3.37 Let $X = \{R \sqsubseteq S_1, R \sqsubseteq S_2\}$, $Y = \{S_1 \sqsubseteq S_3\}$ and $Z = \{S_2 \sqsubseteq S_3\}$. Since $Cn(Y) \cap Cn(Z) = Cn(\emptyset))$, we have that $R \sqsubseteq S_3 \notin Cn(X \cup (Cn(Y) \cap Cn(Z)))$. However, $R \sqsubseteq S_3 \in Cn(X \cup Y) \cap Cn(X \cup Z)$ We conclude that $\mathcal{ALCH}$ is also not distributive.

Flouris studied decomposability of DLs in this thesis [Flo06]. Two of his main results are the following:

Proposition 3.38 ([Flo06]) *Any regular DL that admits intersection* $\sqcap$, *value restriction* $\forall$, *and the universal role* U (*i.e. a role* U *such that* $U^{\mathcal{I}} = \Delta^{\mathcal{I}} \times \Delta^{\mathcal{I}}$ *for any interpretation* $\mathcal{I}$) *is decomposable.*

Proof Let $\mathcal{K}$, $\mathcal{A} \in 2^{\mathcal{L}}$ such that K is logically closed, $Cn(\emptyset) \subset Cn(\mathcal{A}) \subset K$ and $\mathcal{A}$ is finitely representable. Since $\langle \mathcal{L}, Cn \rangle$ is regular then without loss of generality we can set $\mathcal{K} = Cn(\{\top \sqsubseteq A_j : j \in J\})$ and there is $\mathcal{A}'$ equivalent to $\mathcal{A}$ such that $\mathcal{A}' = \{\top \sqsubseteq B_i : 1 \leq i \leq n$. Let $B = B_1 \sqcap \cdots \sqcap B_n$. It is not difficult to verify that $\mathcal{A}'' = \{\top \sqsubseteq B\}$ is equivalent to $\mathcal{A}'$.

---

[3] In some proofs in this section different types of letter will be used to distinguish concepts from sets of sentences.

Let $\mathcal{Z} = \{B \sqsubseteq A_j : j \in J\}$. Of course $Cn(\mathcal{A}'' \cup \mathcal{Z}) = Cn(\mathcal{A} \cup \mathcal{Z}) = \mathcal{K}$. It is also trivial to check that $Cn(\mathcal{Z}) \subseteq \mathcal{K}$. The only thing left to prove is that $Cn(\mathcal{Z}) \neq \mathcal{K}$. Suppose that $\mathcal{K} \subseteq Cn(\mathcal{Z})$, then $Cn(\mathcal{A}) \subseteq Cn(\mathcal{Z})$. Since $Cn(\emptyset) \subset Cn(\mathcal{A})$ there is an interpretation $\mathcal{I}$ such that $B^{\mathcal{I}} \neq \Delta^{\mathcal{I}}$. For this interpretation $\mathcal{I}$ we have that $(\forall U.B)^{\mathcal{I}} = \emptyset$. It follows that $\models_{\mathcal{I}} \mathcal{Z}$ and $\not\models_{\mathcal{I}} \mathcal{A}$. Hence $Cn(\mathcal{Z}) \neq \mathcal{K}$. $\quad\square$

Proposition 3.39 [Flo06] *If a DL admits role hierarchy and at least one of the constructors:*

- *value restriction* ($\forall$),
- *existential restriction* ($\exists$) *or*
- *number restriction* ($\leq_n$).

and does not admit role constructors then it is not decomposable.

Proof Let $\mathcal{K} = Cn(\{R \sqsubseteq S\})$ and $A = \{x \in K : Cn(x) \sqsubset K\}$. By lemma 2.9 we only need to prove that (1) $Cn(A) \neq Cn(\emptyset)$ and (2) $Cn(A) \subset K$.

1. Since the logic admits value restriction, existential restriction, or number restriction, at least one of the following non-tautological sentences is in $Cn(A)$:

$$\begin{aligned}
\exists R.B &\sqsubseteq \exists S.B \\
\forall R.B &\sqsubseteq \forall S.B \\
\leq_n R.B &\sqsubseteq \leq_n R.B
\end{aligned}$$

2. Consider the following interpretation $\mathcal{I}$:

$$\begin{aligned}
\Delta^{\mathcal{I}} &= \{a_1, a_2, b_1, b_2\} \\
B^{\mathcal{I}} &= \emptyset \quad \text{for all concept } B \\
R'^{\mathscr{I}} &= \emptyset \quad \text{for all roles } R' \text{ different from } R \text{ and } S \\
R^{\mathcal{I}} &= \{(a_1, b_1), (b_1, a_1), (a_2, b_2), (b_2, a_2)\} \\
S^{\mathcal{I}} &= \{(a_1, b_2), (b_2, a_1), (a_2, b_1), (b_1, a_2)\}
\end{aligned}$$

Since the logic does not admit role constructors then every relationship between roles are tautological. Thus, it is not difficult to verify that this interpretation satisfies A and do not satisfy K. Hence $Cn(A) \neq K$. $\quad\square$

Two roles R, S are *not related* in a DL $\langle \mathscr{L}, Cn \rangle$ iff $\not\models R \sqsubseteq$ and $\not\models S \sqsubseteq R$.
Proposition 3.40 *If a DL $\langle \mathscr{L}, Cn \rangle$ is regular admits value restriction and has an infinite number of unrelated roles in its signature then $\langle \mathscr{L}, Cn \rangle$ is not complemented.*

Proof Let A be a concept such that $Cn(\top \sqsubseteq A) \neq \mathscr{L}$. Since the logic is regular, the complements of $Cn(\top \sqsubseteq A)$ must be of the form $Cn(\top \sqsubseteq B)$. Suppose w.l.g. that $Cn(\top \sqsubseteq B)$ is a complement of $Cn(\top \sqsubseteq A)$. $Cn(\top \sqsubseteq B) \neq Cn(\emptyset)$, otherwise since $Cn(\top \sqsubseteq A) \cup Cn(\top \sqsubseteq B) = \mathscr{L}$ then $Cn(\top \sqsubseteq A) = \mathscr{L}$ which contradicts the hypothesis. Since $Cn(\top \sqsubseteq A) \neq Cn(\emptyset)$, there is an interpretation $\mathcal{I}$ such that $A^{\mathcal{I}} \neq \Delta^{\mathcal{I}}$. We split the rest of the proof in two parts.

First assume that for every interpretation $\mathcal{I}$ we have that if $A^{\mathcal{I}} \neq \Delta^{\mathcal{I}}$ then $B^{\mathcal{I}} = \Delta^{\mathcal{I}}$. In the case $A \sqsubseteq B \in Cn(\emptyset)$ and $\top B \in Cn(\top \sqsubseteq A) \cap Cn(\top \sqsubseteq B)$. Since

Table 3.2 Logic properties

Logics	Distributive	Complemented	Decomposable	Compact
CPL	yes	yes	yes	yes
Horn	no	no	no	yes
Int	yes	no	no	yes
$\mathcal{ALC} - \mathcal{ALCIQ}$ (no ABox)	yes	no	yes	yes
$\mathcal{S} - \mathcal{SIQ}$ (no ABox)	yes	no	yes	yes
$\mathcal{ALCO} - \mathcal{ALCOIQ}$	yes	no	yes	yes
$\mathcal{SO} - \mathcal{SOIQ}$	yes	no	yes	yes
$\mathcal{ALCH} - \mathcal{ALCHOIQ}$	no	no	no	yes
$\mathcal{SH}$ and $\mathcal{SROIQ}$	no	no	no	yes
$\mathcal{SHIF}$(D) and $\mathcal{SHOIN}$(D)	no	no	no	yes

$\top \sqsubseteq B \notin Cn(\emptyset)$ we have that $Cn(\top \sqsubseteq A) \cap Cn(\top \sqsubseteq B) \neq Cn(\emptyset)$ which contradicts the fact that $Cn(\top \sqsubseteq B)$ is a complement for $Cn(\top \sqsubseteq A)$.

Now assume that there is an interpretation $\mathcal{I}$ such that $A^{\mathcal{I}} \neq \Delta^{\mathcal{I}}$ and $B^{\mathcal{I}} \neq \Delta^{\mathcal{I}}$. Let R be a role which is unrelated with every role that occurs in A and in B. Since $A^{\mathcal{I}} \neq \Delta^{\mathcal{I}}$ and $B \neq \Delta^{\mathcal{I}}$ there are $a, b \in \Delta^{\mathcal{I}}$ such that $a \notin A$ and $b \notin B^{\mathcal{I}}$. Let $(a, b) \in R^{\mathcal{I}}$ then $a \notin \forall R.B$. It follows that $a \notin A^{\mathcal{I}} \cup (\forall R.B)^{\mathcal{I}}$. Hence, $\top \sqsubseteq A \sqcup \forall R.B \notin Cn(\emptyset)$, but $\top \sqsubseteq A \sqcup \forall R.B \in Cn(\top \sqsubseteq A) \cap Cn(\top \sqsubseteq B)$. □

The literature presents two definitions of consistency in DLs. A concept A is called *satisfiable* iff there is at least one interpretation $\mathcal{I}$ such that $A^{\mathcal{I}} \neq \emptyset$. The concept $A \sqcap \neg A$, for example, is not satisfiable since $(A \sqcap \neg A)^{\mathcal{I}} = \emptyset$ for every interpretation $\mathcal{I}$. The presence of unsatisfiable concepts in a belief set normally indicates a modeling error. A set of sentence that implies an unsatisfiable concept is considered inconsistent.

The second type of inconsistency is much more critical. It indicates that there is no possible interpretation for the base. A set of sentences B is inconsistent if there is no interpretation $\mathcal{I}$ such that $\vDash_{\mathcal{I}} B$.

3.5 Conclusion

Although the abstraction of consequence operator is enough for most results in the future chapter, it is useful to present examples of concrete logics. For this reason in this chapter several logics was presented, namely, Classical Propositional Logic, Intuitionistic Logic, Horn Logic, and some Description Logics.

CPL was presented as a canonical example of well-behaved logic, i.e., logic that satisfies the AGM assumptions. Other logics such as modal logics are also well behaved, but were omitted here (see [Che80] for a good introduction in modal logics). We presented both axiomatic system and semantics of CPL. Our approach followed [Eps90], but there are lots of other good references for CPL in the literature.

Intuitionistic logic was developed by Heyting (see [Min00] for more details about Intuitionistic Logic). It was presented here for mainly two reasons: it is well known by logicians and it has certain interesting properties such as being distributive, but not decomposable.

Horn Logic and Description Logics were presented because they have computational interested. Horn Logic is widely used in Artificial Intelligence application because of its low computational complexity and Description Logics are used as theoretical framework for ontologies on the Web. For a more detailed presentation of Description Logics and its applications see [BCM+03].

The main aspects of these logics that we need to focus are the properties that their consequence operator satisfy. Results on future chapters are applicable to a logic depending on the properties of its consequence operator. We summed up the results concerning the properties that the consequence operator of each logic satisfy in Table 3.2 which should be used as reference.

References

[ACKZ09]   Alessandro Artale, Diego Calvanese, Roman Kontchakov, and Michael Zakharyaschev. The DL-lite family and relations. Technical Report BBKCS-09-03, School of Computer Science and Information Systems, Birbeck College, 2009.

[AW97]   G. Antoniou and M.A. Williams. *Nonmonotonic reasoning*. Artificial intelligence. MIT Press, 1997.

[BBL08]   Franz Baader, Sebastian Brandt, and Carsten Lutz. Pushing the el envelope further. In Kendall Clark and Peter F. Patel-Schneider, editors, *Proceedings of the OWLED 2008 DC Workshop on OWL: Experiences and Directions, 2008*.

[BCM+03]   Franz Baader, Diego Calvanese, Deborah McGuinness, Daniele Nardi, and Peter Patel-Schneider, editors. *The Description Logic Handbook*. Cambridge University Press, 2003.

[CE06]   Walter Carnielli and Richard L. Epstein. *Computabilidade Funções Computáveis, Lógica e os Fundamentos da Matemática*. UNESP, 2006.

[Che80]   B.F. Chellas. *Modal logic: an introduction*. Cambridge University Press, 1980.

[EH86]   E Allen Emerson and Joseph Y Halpern. Sometimes and not never revisited: on branching versus linear time temporal logic. *Journal of the ACM*, 33(1):151–178, 1986.

[Eps90]   R. L. Epstein. *The Semantics Foundations of Logic. Volume 1: Propositional Logics*. Kluwer, Dordrecht, Netherlands, 1990.

[Flo06]   Giorgos Flouris. *On Belief Change and Ontology Evolution*. PhD thesis, University of Crete, 2006.

[Gir87]   Jean-Yves Girard. Linear logic. *Theoretical Computer Science*, 50:1–102, 1987.

[Gru93]   T. R. Gruber. Towards Principles for the Design of Ontologies Used for Knowledge Sharing. In N. Guarino and R. Poli, editors, *Formal Ontology in Conceptual Analysis and Knowledge Representation*, Deventer, Netherlands, 1993. Kluwer Academic Publishers.

[HM01]   Volker Haarslev and Ralf Möller. Racer system description. *In Proceedings of the International Joint Conference on Automated Reasoning (IJCAR 2001)*, volume 2083 of Lecture Notes in Artificial Intelligence, pages 701–705. Springer, 2001.

[Hor98]   Ian Horrocks. Using an expressive description logic: FaCT or fiction? In A. G. Cohn, L. Schubert, and S. C. Shapiro, editors, *Proceedings of the 6th International*

Conference on the Principles of Knowledge Representation and Reasoning (KR'98), pages 636–647, Trento, Italia, Junho, 2-5 1998. Morgan Kaufmann.

[LB87] Hector J. Levesque and Ronald J. Brachman. Expressiveness and tractability in knowledge representation and reasoning. *Computational Intelligence*, 3:78–93, 1987.

[MGH+08] Boris Motik, Bernardo Cuenca Grau, Ian Horrocks, Zhe Wu, Achille Fokoue, and Carsten Lutz. OWL 2 web ontology language: Profiles. World Wide Web Consortium, Working Draft, Dezembro 2008. Available at http://www.w3.org/TR/2008/WD-owl2-profiles-20081202/.

[Min81] Marvin Minsky. A framework for representing knowledge. In John Haugeland, editor, Mind Design: *Philosophy, Psychology, Artificial Intelligence*, pages 95–128. MIT Press, Cambridge, MA, 1981.

[Min00] Grigori Mints. *A short introduction to intuitionistic logic*. Kluwer Academic Publishers, Norwell, MA, USA, 2000.

[MSH07] Boris Motik, Rob Shearer, and Ian Horrocks. Optimized Reasoning in Description Logics using Hypertableaux. In Frank Pfenning, editor, *Proceedings of the 21st Conference on Automated Deduction (CADE- 21)*, volume 4603 of *LNAI*, pages 67–83, Bremen, Germany, July 17-20 2007. Springer.

[OWL09] W3C OWL Working Group. *OWL 2 Web Ontology Language*: Document Overview. W3C Recommendation, 27 October 2009. Available at http://www.w3.org/TR/owl2-overview/.

[Qui67] Ross M. Quillian. Word concepts: A theory and simulation of some basic semantic capabilities. *Behavioral Science*, 12:410–430, 1967.

[RRP96] A. L. Rector, J. E. Rogers, and P. A. Pole. The GALEN high level ontology. In *Proceedings of Medical Informatics Europe 96*, pages 174–178. IOS Press, Janeiro 1996.

[SR04] Kent A. Spackman and Guillermo Reynoso. Examining SNOMED from the perspective of formal ontological principles: Some preliminary analysis and observations . In Udo Hahn, editor, *1st International Workshop on Formal Biomedical Knowledge Reprsentation (KR-MED)*, volume 102 of CEUR Workshop Proceedings, pages 72–80, Whistler, BC, Canada, Junho, 1 2004. CEUR-WS.org.

[TH06] D. Tsarkov and I. Horrocks. Fact++ description logic reasoner: System description. In *Proceedings of the International Joint Conference on Automated Reasoning (IJCAR 2006)*, volume 4130 of *Lecture Notes in Artificial Intelligence*, pages 292–297. Springer, 2006.

Chapter 4
Classical Belief Revision

Abstract Belief revision is the subarea of knowledge representation which studies the dynamics of epistemic states of an agent. Belief systems are composed of three pieces: a representation of epistemic states, a set of epistemic attitudes, and types of belief change. Two specific belief systems are presented in this chapter, namely, AGM theory and belief base theory

Keywords Belief revision · Partial meet contraction · AGM theory · Belief base theory · Belief systems

This chapter introduces the main concepts and results about classical belief revision theory. Next sections define generic belief systems following [Gar88]. Section 4.2 presents a belief system where epistemic states are represented as a set closed under logical consequence. This approach is called *AGM theory* in honor to the authors of [AGM85]. Section 4.3 presents a belief system where epistemic states are represented as arbitrary sets of sentences. This approach distinguishes implicit from explicit beliefs and is called *belief base theory*.

4.1 Belief Systems

An agent's epistemic state is the set of beliefs of the agent at certain moment. Gärdenfors defines it as an idealized representation of the cognitive state of an agent at certain moment [Gar88]. It is possible to avoid making such strong assumption about the cognitive state of the agent and defining the epistemic state as a representation of every belief that can be attributed to it at certain moment, but this is not our main focus here.

There are several ways of modeling the epistemic state of an agent. This work considers only models based on sets of sentences. Probabilistic models, as in [Pea88] for example, would not be considered.

M. M. Ribeiro, *Belief Revision in Non-Classical Logics*,
SpringerBriefs in Computer Science,
DOI: 10.1007/978-1-4471-4186-0_4, © The Author(s) 2013

Sections 4.2 and 4.3 present two ways of representing epistemic states: as sets of sentences logically closed (belief sets) and as arbitrary sets of sentences (belief bases).

Given a model of the epistemic state, an agent may have a series of *epistemic attitudes* toward each element of the model. In a probabilistic model, for example, an agent may accept or reject a sentence with certain probability. We will see in Sect. 4.2 that, when modeling the epistemic states as a belief set, the agent may have three attitudes before a sentence: accept, reject, or indeterminate (do not have opinion). When modeling the epistemic state as a belief base, on the other hand, the agent may have four attitudes: explicitly accept, implicitly accept, reject, or indeterminate.

Belief revision studies the dynamics of epistemic states of an agent, i.e., how agents change their attitude toward an element of the model. Following Gärdenfors [Gär88], these changes are fired by an external trigger called *epistemic input*. The form of this input is irrelevant. It can be uttered sentence, a smell, a noise, etc. For our purposes the only thing that matters is the effect of this input in the epistemic state of the agent.

An epistemic input can lead to several kinds of epistemic change. We consider three types of *belief change*, also called *operations*, triggered by the epistemic input:

expansion: makes the agent to accept a new sentence.
revision: makes the agent to accept a new sentence in a consistent manner.
contraction: makes the agent to abandon the belief in a sentence.

Picture an agent in certain epistemic state. Imagine now that some external trigger make it change its mind, i.e., drives it to a different epistemic state. The belief change is the function that given an epistemic state and an input returns the new epistemic state.

The main challenge of belief revision is to define *rationality criteria* for belief change. What can one expect from an agent when it changes its beliefs? For example, one such criterion defended by many authors, such as Harman [Har86] for example, is the *minimality criterion*. When an agent changes its mind we expect it not to abandon beliefs unnecessarily.

Sections 4.2.1 and 4.2.2 present a list of rationality postulates for contraction and for revision in belief sets. Furthermore, these sections also present rationality postulates for operations in belief bases.

Besides presenting the rationality postulates, we need to show how to construct each of these operations. A set of postulates *characterizes* a construction if, on the one hand, the construction satisfies each postulate and, on the other hand, any operation that satisfies the postulates can be obtained by the construction. This equivalence between rationality postulates and certain construction is a central result in belief revision theory. For each operation defined in this chapter a set of rationality postulate, a construction, and the *representation theorems* showing their equivalence are presented.

A *belief system* is a model for epistemic states together with a definition of epistemic attitudes, a class of epistemic inputs, and a set of operations.

4.2 AGM Theory

AGM theory [AGM85, Gar88] studies changes in epistemic states modeled as logically closed sets of sentences. The set of sentences that represents the agent's epistemic state is called *belief set*. Let us assume for the rest of this section that the agents reason according to certain logic $\langle \mathcal{L}, Cn \rangle$ and that this logic satisfies the AGM assumptions. Belief sets are sets of sentences closed under logic consequence, i.e., given an underlying logic $\langle \mathcal{L}, Cn \rangle$, a belief set K satisfies $K = Cn(K)$ or equivalently $K \in \mathbb{K}_{\mathcal{L}}$.

Three types of epistemic attitudes toward a sentence are considered. Let K be the belief set of an agent. A sentence α can be:

accepted: if $\alpha \in K$
rejected: if $K \cup \{\alpha\}$ is inconsistent
indeterminate: if $\alpha \notin K$ and $K \cup \{\alpha\}$ is consistent.

Three types of belief change, called operations, in belief sets are considered: expansion, contraction, and revision. Let K be a belief set in a logic $\langle \mathcal{L}, Cn \rangle$, the symbols $K + \alpha$, $K - \alpha$, and $K * \alpha$ represent the belief set of the agent after an expansion by α, a contraction by α and a revision by α, respectively.

Expansion is certainly the simplest of those operations. An expansion consists simply of the acceptance of a sentence α. Expansion can easily be achieved using the following formula:

$$K + \alpha = Cn(K \cup \{\alpha\})$$

Revision consists of the consistent acceptance of a sentence α. Besides guarantying that the input is accepted ($\alpha \in K * \alpha$) and that the resulting belief set is consistent ($K * \alpha$ is consistent), revision should guaranty that the change is somehow minimal. Notice that revision is way more complicated than expansion, since it involves "extra-logical" components.

Consider Example 4.1. To accommodate the belief that Socrates is immortal Aristotle has to abandon some of its old beliefs. However, there is no logical criterion to decide which beliefs should be abandoned. For this reason, revision cannot be defined in the same way that expansion was. Revision is defined via a set of rationality postulates (Sect. 4.2.2).

Contraction consists of the removal of a sentence α from the belief set K. Besides guarantying that the input is indeterminate in the new belief set ($\alpha \notin K - \alpha$) contraction should guaranty that $K - \alpha$ is a belief set and that the change is somehow minimal. Contraction also depends on "extra-logical" factors. Its postulates are presented in the following subsection.

4.2.1 AGM Contraction

In this section contraction will be defined through rationality postulates called *AGM postulates for contraction*. After that, a construction for contraction called *partial meet* is presented. Both postulates and construction were originally presented in [AGM85]. This seminal work also presented the representation theorem relating the construction with the postulates. This theorem is also presented in this section.

AGM Postulates
When an agent questions the veracity of some of its beliefs, he evaluates a contraction, i.e., contraction corresponds with the action of open minding.

Example 4.1 I believe that every man is a mortal $man \rightarrow mortal$, but Hannah believes that this is not the case. I may contract $man \rightarrow mortal$ for the sake of argumentation.

Contraction is an operation defined over the actual belief set K that given a sentence α as input returns a new belief set $K - \alpha$ that does not contain the *input* ($\alpha \notin K - \alpha$).

Since contraction depends on extra-logical components, we define it indirectly via a set of rationality postulates. These postulates are known as *AGM-postulates for contraction*.

The first of these postulates is called *closure*. Closure guarantees that the result $K - \alpha$ of a contraction is a belief set, i.e.,

(**closure**) $K - \alpha = Cn(K - \alpha)$

A successful contraction should remove the input α from the original belief set K. This is possible for every input α except when α is a tautology, i.e., when $\alpha \in Cn(\emptyset)$. If α is tautological, removing α from K would imply that $K - \alpha$ is not closed, i.e., it would violate closure. Hence, the second AGM postulate is stated as follows:

(**success**) If $\alpha \notin Cn(\emptyset)$ then $\alpha \notin Cn(K - \alpha)$.

Furthermore, in an epistemic change only the content of the sentences should matter, not its syntactic form. Hence, if α and β are equivalent then contracting either of them from K should give us the same resulting belief set.

(**extensionality**) If $Cn(\alpha) = Cn(\beta)$ then $K - \alpha = K - \beta$.

The following three postulates captures minimality aspects of contraction. *Inclusion* states that to perform a contraction, there is no need of adding sentences to K.

(**inclusion**) $K - \alpha \subseteq K$

Vacuity deals with an extreme case. If the input α is already not an element of K, then the minimal thing to do is to leave K unchanged.

(**vacuity**) If $\alpha \notin K$ then $K - \alpha = K$.

An operation that satisfies the postulates presented so far (i.e., success, closure, extensionality, inclusion and vacuity) is called withdrawal. Withdrawal was first presented in [Mak87] as an alternative to contraction. Hansson argues that this operation does not properly guaranty minimality and illustrates this point with an example [Han91] Consider the following function – over K :

$$K - \alpha = \begin{cases} K & \text{if } \alpha \notin Cn(K) \\ Cn(\emptyset) & \text{otherwise.} \end{cases}$$

Although very unintuitive, this function is a valid withdrawal for any α.

The last and most controversial AGM-postulate tries to capture this notion of minimality. This postulate, called *recovery*, guarantees that if α is contracted from K then the new belief set $K - \alpha$ should retain enough information from K so that if $K - \alpha$ is expanded by α it recovers every sentence from K.

(**recovery**) $K \subseteq (K - \alpha) + \alpha$

Although being a very important minimality criterion, several authors criticize recovery [Mak87, Han91, FPA06]. An important part of this work is devoted to analyze this postulate in the context of non-classical logics (see Chap. 5).

The six postulates presented so far are called *basic AGM-postulates for contraction* or just *AGM-postulates for contraction* [AGM85]:

(**closure**) $K - \alpha = Cn(K - \alpha)$
(**success**) If $\alpha \notin Cn(\emptyset)$ then $\alpha \notin K - \alpha$
(**inclusion**) $K - \alpha \subseteq K$
(**vacuity**) If $\alpha \notin K$ then $K - \alpha = K$
(**recovery**) $K \subseteq (K - \alpha) + \alpha$.
(**extensionality**) If $Cn(\alpha) = Cn(\beta)$ then $K - \alpha = K - \beta$

In [AGM85], the authors present other two postulates called auxiliaries. These postulates won't be discussed in the present work.

Partial Meet Contraction

Belief revision literature presents several constructions for contraction that satisfies the AGM postulates (see [Gar88]). In this section it is presented one of these constructions called *partial meet contraction*.

Since we are interested in minimal modifications of the belief set, it is worth investigating the maximal subsets of K that fail to imply the input α. The set of all these subsets of K is called *remainder set*:

Definition 4.2 (*remainder set*) [AGM85] Let B be a set of sentences and α a sentence in a logic $\langle \mathscr{L}, Cn \rangle$. The remainder set $B \perp \alpha$ is the set whose elements X are maximal subsets of B that does not imply α, i.e., $X \in B \perp \alpha$ iff:

1. $X \subseteq B (X$ is a subset of B).
2. $\alpha \notin Cn(B)$(that doesn't imply α).
3. if $X \subset X' \subseteq B$ then $\alpha \in Cn(B')$(and is maximal).

The following are lemmas about remainder set that are needed for future results:

Lemma 4.3 (upper bound property) [AM81] *For every set of sentences B, every subset X of B and every sentence α in a compact logic $\langle \mathcal{L}, Cn \rangle$ for which $\alpha \notin Cn(X)$, there is a X' such that $X \subseteq X'$ and $X' \in B \perp \alpha$.*

Proof We adapted the proof of Lindembaum property in [Woj88]. First arrange the sentences of B into a sequence $\beta_1, \beta_2, \ldots$.
Now let $X_0 = X$ and for each $i \geq 1$ define X_i as follows:

$$X_i = \begin{cases} X_{i-1} & \text{if } \alpha \in Cn(X_{i-1} \cup \{\beta_i\}) \\ X_{i-1} \cup \{\beta_i\} & \text{otherwise.} \end{cases}$$

By construction, for every i we have that $\alpha \notin Cn(X_i)$. Now let $X' = \bigcup_i X_i$. It is trivial to verify that $X \subseteq X' \subseteq B$. Moreover if $\beta \in B$ and $\beta \notin X'$ then, by construction, $\alpha \in X' \cup \{\beta\}$.
By compactness, if $\alpha \in Cn(X')$ then $\alpha \in Cn(X'')$ for some finite $X'' \subseteq B$. It would follow that $\alpha \in Cn(X_i)$ for some i which is a contradiction.   □

Lemma 4.4 *Let B be a set of sentences and α be a sentence in a compact logic $\langle \mathcal{L}, Cn \rangle$. $B \perp \alpha = \emptyset$ if and only if $\alpha \in Cn(\emptyset)$.*

Proof ($\Rightarrow$) If $\alpha \notin Cn(\emptyset)$ then by upper bound property (Lemma 4.3) there is X such that $Cn(\emptyset) \subseteq X \in B \perp \alpha$. Hence $B \perp \alpha \neq \emptyset$.
($\Leftarrow$) If $\alpha \in Cn(\emptyset)$ then by *monotonicity* $\alpha \in X$ for every $X \subseteq B$. Hence $B \perp \alpha = \emptyset$.   □

Lemma 4.5 *Let K be a belief set and α a sentence in a logic $\langle \mathcal{L}, Cn \rangle$. If $X \in K \perp \alpha$ then X is closed under logical consequence i.e. $X = Cn(X)$.*

Proof If $\beta \in Cn(X)$ then $\alpha \notin Cn(X \cup \{\beta\})$ and, since X is maximal, $\beta \in X$.   □

Lemma 4.6 *Let B be a set of sentences and α and β be sentences in a logic $\langle \mathcal{L}, Cn \rangle$. $B \perp \alpha = B \perp \beta$ if and only if for every subset B' of B it holds that $\alpha \in Cn(B')$ iff $\beta \in Cn(B')$*

Proof ($\Rightarrow$) Let $B' \subseteq B$ and $\alpha \in Cn(B')$. Suppose that $\beta \notin Cn(B')$ then by *upper bound property* (Lemma 4.3) there is X such that $B' \subseteq X \in B \perp \beta$. Since $\alpha \in Cn(B')$, by *monotonicity* $\alpha \in Cn(X)$ and, hence, $X \notin B \perp \alpha$.
($\Leftarrow$) This holds trivially.   □

Lemma 4.7 *Let K be a belief set and α and β be sentences in a compact logic $\langle \mathcal{L}, Cn \rangle$ then the following holds:*

1. *If $Cn(\alpha) = Cn(\beta)$ then $K \perp \alpha = K \perp \beta$.*
2. *If $\alpha \in K$ and $K \perp \alpha = K \perp \beta$ then $Cn(\alpha) = Cn(\beta)$.*

Proof

1. If $K' \subseteq K$ then, since $Cn(\alpha) = Cn(\beta)$, $\alpha \in Cn(K')$ iff $\beta \in Cn(K')$. It follows, by the previous lemma that $K \perp \alpha = K \perp \beta$.
2. If $\alpha \in K$ then $Cn(\alpha) \in K$. By the previous lemma, $\beta \in Cn(\alpha)$. Analogously, if $\beta \in K$ then $\alpha \in Cn(\beta)$ and we conclude that $Cn(\alpha) = Cn(\beta)$, by *monotonicity* and *idempotence*.

   If $\beta \notin K$ then $K \perp \beta = \{K\}$ and, since $K \perp \alpha = K \perp \beta$, we have that $\alpha \notin K$.

   We conclude that if $\alpha \in K$ and $K \perp \alpha = K \perp \beta$ then $Cn(\alpha) = Cn(\beta)$. $\qquad\square$

Consider a function γ that selects certain elements of $K \perp \alpha$ when possible and returns $\{K\}$ otherwise. This function is called *selection function* and it intuitively selects the sets of sentences that are more entrenched in the agents beliefs. Formally:

Definition 4.8 (*selection function*) [AGM85] Let K be a belief set. A *selection function* for K is a function γ such that for every α :

1. $\emptyset \neq \gamma(K \perp \alpha) \subseteq K \perp \alpha$ if $K \perp \alpha \neq \emptyset$.
2. $\gamma(K \perp \alpha) = \{K\}$ otherwise.

Partial meet contraction consists of the intersection of the sets selected by γ. Formally, let $K \perp \alpha$ be a remainder set and γ a selection function for it. Partial meet contraction $-_\gamma$ is defined as:

$$K -_\gamma \alpha = \bigcap \gamma(K \perp \alpha)$$

An operation $-$ is a partial meet contraction iff there is a selection function γ such that for every sentence α we have that $K - \alpha = K -_\gamma \alpha$.

The main result in [AGM85] shows that AGM postulates fully characterizes partial meet contraction i.e. every partial meet contraction satisfies AGM postulates and every AGM contraction is a partial meet contraction. In other words an operation $-$ over a belief set K is an AGM contraction iff it is a partial meet contraction. This result is known in the literature as the *representation theorem for partial meet contraction*:

Representation Theorem 4.9 [AGM85] *Let $\langle \mathscr{L}, Cn \rangle$ be a logic that satisfies the AGM assumptions. An operation $-$ over a belief set K is a partial meet contraction iff it satisfies closure, success, inclusion, vacuity, recovery, and extensionality, i.e., the basic AGM postulates.*

Proof (**construction $\Rightarrow$ postulates**)

closure: If $\alpha \notin K$ then $\gamma(K \perp \alpha) = \{K\}$ and $\bigcap \gamma(K \perp \alpha) = K$ which is closed by hypothesis. If $\alpha \in K$ then $\emptyset \neq \gamma(K \perp \alpha) \subseteq K \perp \alpha$ and by Lemma 4.5 every element of $K \perp \alpha$ is logically closed. Since the intersection of closed sets is closed (Lemma 2.1), $\bigcap \gamma(K \perp \alpha)$ is logically closed.

success: Follows directly by Lemma 4.4.

inclusion: Follows directly from construction.

vacuity: If $\alpha \notin K$ then $K \perp \alpha = \{K\}$ and $\bigcap \gamma(K \perp \alpha) = K$.

recovery: If $\alpha \notin K$ then $K - \alpha = K$, hence, $K \subseteq (K - \alpha) + \alpha$. If $\alpha \in K$ then
  let $\beta \in K$ and suppose that $\alpha \to \beta \notin X$ for some $X \in K \perp \alpha$. Since X is
  maximal $\alpha \in Cn(X \cup \{\alpha \to \beta\})$. By *deduction* $(\alpha \to \beta) \to \alpha \in Cn(X)$
  and by *supraclassicality* $\alpha \in Cn(X)$ which contradicts the definition of X.
  It follows that $\alpha \to \beta \in X$ for every $X \in K \perp \alpha$. Hence, $\alpha \to \beta \in \bigcap \gamma(K \perp \alpha)$.
  By *supraclassicality* $\beta \in (\bigcap \gamma(K \perp \alpha)) + \alpha$.

extensionality: If $Cn(\alpha) = Cn(\beta)$ then by Lemma 4.7 $K \perp \alpha = K \perp \beta$, hence,
  $\bigcap \gamma(K \perp \alpha) = \bigcap \gamma(K \perp \beta)$.

(postulates $\Rightarrow$ construction)

Let $-$ be an operator satisfying the AGM postulates for contraction and the selection function γ as follows:

$$\gamma(K \perp \alpha) = \{X \in K \perp \alpha : K - \alpha \subseteq X\} \quad \text{if } K \perp A \neq \emptyset$$
$$= \{K\} \quad \text{otherwise.}$$

We need to prove (1) γ is well defined (i.e., if $K \perp \alpha = K \perp \beta$ then $\gamma(K \perp \alpha) = \gamma(K \perp \beta)$), (2) γ is a selection function (i.e., $\emptyset \neq \gamma(K \perp \alpha) \subseteq K \perp \alpha$ if $K \perp \alpha \neq \emptyset$) and (3) $K - \alpha = \bigcap \gamma(K \perp \alpha)$.

1. If $\alpha \notin K$ then $K \perp \alpha = K \perp \beta = \{K\}$. Hence $\gamma(K \perp \alpha) = \gamma(K \perp \beta) = \{K\}$.
   If $\alpha \in K$ and $K \perp \alpha = K \perp \beta$ then, by Lemma 4.7, $Cn(\alpha) = Cn(\beta)$ and, by
   extensionality $K - \alpha = K - \beta$. It follows that $\gamma(K \perp \alpha) = \gamma(K \perp \beta)$.

2. It follows directly from construction that $\gamma(K \perp \alpha) \subseteq K$. If $K \perp \alpha \neq \emptyset$ then by
   Lemma 4.4 $\alpha \notin Cn(\emptyset)$. Thus, by *success* and *inclusion* $\alpha \notin K - \alpha \subseteq K$. By
   upper bound property (Lemma 4.3) there is X such that $K - \alpha \subseteq X \in K \perp \alpha$.
   Hence, $\gamma(K \perp \alpha) \neq \emptyset$.

3. If $\alpha \notin K$ then $K - \alpha = K$ by *vacuity*. In this case $K \perp \alpha = \{K\}$ and, hence,
   $\bigcap \gamma(K \perp \alpha) = K - \alpha = K$.
   If $\alpha \in Cn(\emptyset)$ then $K - \alpha = K$ by *recovery* and *inclusion*. In this case $K \perp \alpha = \emptyset$,
   by Lemma 4.4. It follows that $\bigcap \gamma(K \perp \alpha) = K - \alpha = K$.
   Now consider the last case where $\alpha \notin Cn(\emptyset)$ and $\alpha \in K$.
   By construction, $K - \alpha \subseteq \bigcap \gamma(K \perp \alpha)$.
   Now let $\beta \notin K - \alpha$ and suppose that $\alpha \in Cn(K - \alpha \cup \{\beta \to \alpha\})$. By *deduction*
   and *closure* $(\beta \to \alpha) \to \alpha \in K - \alpha$. Furthermore, by *recovery* and *closure*
   $\alpha \to \beta \in K - \alpha$. By *supraclassicality* $\beta \in K - \alpha$ which contradicts the
   hypothesis. It follows that $\alpha \notin Cn(K - \alpha \cup \{\beta \to \alpha\})$. Since $\beta \to \alpha \in Cn(\beta)$,
   we have that $\beta \to \alpha \in K$. It follows, by *upper bound property* (Lemma 4.3)
   that there is X such that $K - \alpha \cup \{\beta \to \alpha\} \subseteq X \in K \perp \alpha$ and, of course,
   $\beta \notin X$. Otherwise $\alpha \in X$ which contradicts the definition of X. We conclude that
   $\beta \notin \bigcap \gamma(K \perp \alpha)$. Hence, $\bigcap \gamma(K \perp \alpha) \subseteq K - \alpha$.   □

4.2.2 AGM Revision

AGM revision is an operation $*$ over a belief set K that consistently adds a new sentence α into K. The result of this operation should be a new belief set $K * \alpha$.

The operation $*$ is also defined via a set of rationality postulates. After presenting these postulates we show how to construct the revision. The representation theorem relating the construction to the postulates is presented in the end of the section.

Revision consists in consistently adding a sentence α into the belief set K. Hence, we have the following postulates:

(**success**) $\alpha \in K * \alpha$

(**consistency**) If α is consistent then $K * \alpha$ is consistent.

Some rationality postulates for revision are analogous with certain postulates for contraction. In these cases we follow the tradition and use the same names for both postulates.

(**closure**) $K * \alpha = Cn(K * \alpha)$

(**inclusion**) $K * \alpha \subseteq K + \alpha$

(*extensionality*) If $Cn(\alpha) = Cn(\beta)$ then $K * \alpha = K * \beta$

(**vacuity**) If $K + \alpha$ is consistent then $K * \alpha = K + \alpha$.

The following postulate, although presented in the original paper [AGM85], had not been used in more recent works:

(**Harper identity**) $K * \neg \alpha \cap K = K - \alpha$ for some AGM contraction – over K.

Makinson [Mak87] argues that this statement should not be used as a postulate, but rather as a way to construct a contraction given a revision. The AGM postulates for revision as defined in [Gar88] are:

(**closure**) $K * \alpha = Cn(K * \alpha)$.

(**success**) $\alpha \in K * \alpha$.

(**inclusion**) $K * \alpha \subseteq K + \alpha$.

(**vacuity**) If $K + \alpha$ is consistent then $K * \alpha = K + \alpha$

(**consistency**) If α is consistent then $K * \alpha$ is consistent.

(**extensionality**) If $Cn(\alpha) = Cn(\beta)$ then $K * \alpha = K * \beta$.

Levi Identity

The canonical way to construct AGM revision is using the AGM contraction – via Levi identity.

Let – be an AGM contraction over K. The revision $*$ over K defined using the Levi identity is the following:

$$K * \alpha = (K - \neg \alpha) + \alpha$$

The revision defined via Levi identity satisfies all the AGM postulates. In fact, if – is a withdrawal operation, i.e. if – satisfies all AGM postulates for contraction but recovery, then $*$ defined via Levi identity satisfies the AGM postulates for revision.

Theorem 4.10 [Mak87] *Let K be a belief set and α be a sentence in a logic $\langle \mathscr{L}, Cn \rangle$ that satisfies the AGM assumptions and let – be a withdrawal operation. The revision* *operation * over K defined via Levi identity using – satisfies the AGM postulates for revision.*

Proof success and closure: Follows directly from the construction.

   inclusion: By *inclusion of contraction* $K - \neg\alpha \subseteq K$. Hence, $(K - \neg\alpha) + \alpha \subseteq K + \alpha$. *extensionality:* If $Cn(\alpha) = Cn(\beta)$ then $Cn(\neg\alpha) = Cn(\neg\beta)$. By *extensionality of contraction* $K - \neg\alpha = K - \neg\beta$. Hence, $(K - \neg\alpha) + \alpha = (K - \neg\beta) + \beta$. *vacuity:* If $K + \alpha$ is consistent then $\neg\alpha \notin K$. It follows by *vacuity of contraction* that $K - \neg\alpha = K$. Hence, $(K - \neg\alpha) + \alpha = K + \alpha$. *consistency:* If α is consistent then $\neg\alpha \notin Cn(\emptyset)$. It follows by *success of contraction* that $\neg\alpha \notin K - \neg\alpha$. By *$\alpha$-local non-contravention* that $\neg\alpha \notin (K - \neg\alpha) + \alpha$. Hence, $(K - \neg\alpha) + \alpha$ is consistent.   □

A similar result can be proved using the Harper identity. If $*$ is an AGM revision then the contraction – defined via Harper identity (i.e., $K - \alpha = K * \neg\alpha \cap K$) satisfies the AGM postulates for contraction.

Theorem 4.11 [Mak87] *Let K be a belief set and α be a sentence in a logic $\langle \mathcal{L}, Cn \rangle$ that satisfies the AGM assumptions and let $*$ be an AGM revision. The contraction operation – over K defined via Harper identity using – satisfies the AGM postulates for contraction.*

Proof closure: Follows from the *closure of revision* and the fact that the intersection of closed sets is closed. *success:* If $\alpha \notin Cn(\emptyset)$ then $\neg\alpha$ is consistent. By *consistency of revision* $K * \neg\alpha$ is consistent and, by *success of revision* $\neg\alpha \in K * \neg\alpha$. It follows that $\alpha \notin K * \neg\alpha$. Hence, $\alpha \notin K * \neg\alpha \cap K$. *inclusion:* Follows directly from construction. *vacuity:* If $\alpha \notin K$ then $K + \neg\alpha$ is consistent and, by *vacuity of revision*, $K * \neg\alpha = K + \neg\alpha$. Hence, $K \cap K * \neg\alpha = K$. *extensionality:* If $Cn(\alpha) = Cn(\beta)$ then $Cn(\neg\alpha) = Cn(\neg\beta)$ and, by *extensionality of revision* $K * \neg\alpha = K * \neg\beta$. We conclude that $K \cap K * \neg\alpha = K \cap K * \neg\beta$. *recovery:* If $\alpha \notin K$ then $K \cap K * \neg\alpha = K$. Hence, $K \subseteq (K - \alpha) + \alpha$ If $\alpha \in K$ and $\beta \in K$ then $\alpha \to \beta \in K$. Furthermore, by *success of revision* $\neg\alpha \in K * \neg\alpha$. It follows that $(K * \neg\alpha) + \alpha$ is not consistent and, by *supraclassicality*, trivial.

Hence, $\beta \in (K * \neg\alpha) + \alpha$ and, by *deduction* and *closure of revision*, $\alpha \to \beta \in K * \neg\alpha$. We conclude that $\alpha \to \beta \in K \cap K * \alpha$ and, hence, $\beta \in (K \cap K * \neg\alpha) + \alpha$ □

Theorems 4.10 show that any method of construction for the AGM contraction – can be used to define the revision $*$. As a concrete example take a partial meet contraction $-_\gamma$. The revision defined using Levi identity $K *_\gamma \alpha = (K -_\gamma \neg\alpha) + \alpha$ is called partial meet revision. Partial meet revision is fully characterized by AGM postulates for revision.

Representation Theorem 4.12 [1] *Let K be a belief set in $\mathcal{L}$ and let $\alpha \in \mathcal{L}$. The operation $*$ over K is a partial meet revision iff $*$ satisfies closure, success, inclusion, vacuity, consistency, and extensionality, i.e., the AGM postulates for revision.*

---

[1]   This result was proved in [Han99], but the proof presented here is new.

Proof (**construction** ⇒ **postulates**)

success, closure, and inclusion: follows directly from construction.

vacuity: If $K + \alpha$ is consistent then $\neg\alpha \notin K$. It follows that $K\bot\neg\alpha = \{K\}$ and $(\bigcap \gamma(K\bot\alpha)) + \alpha = K + \alpha$.

consistency: If α is consistent then $\neg\alpha \notin Cn(\emptyset)$. By *upper bound property* (Lemma 4.3) there is $X \in K\bot\neg\alpha$. It follows that $\alpha \notin \bigcap \gamma(K\bot\neg\alpha)$ and, by α-*local non contravention*, $\alpha \notin (\bigcap \gamma(K\bot\neg\alpha)) + \alpha$.

extensionality: If $Cn(\alpha) = Cn(\beta)$ then $Cn(\neg\alpha) = Cn(\neg\beta)$. By Lemma 4.7, $K\bot\neg\alpha = K\bot\neg\beta$, hence, $(\bigcap \gamma(K\bot\neg\alpha)) + \alpha = (\bigcap \gamma(K\bot\neg\beta)) + \beta$

(**postulates** ⇒ **construction**)

Let $*$ be an operation over K that satisfies the AGM postulates for revision and let the selection function γ be defined as follows:

$$\gamma(K\bot\alpha) = \{X \in K\bot\alpha : K \cap K * \neg\alpha \subseteq X\} \text{ if } K\bot A \neq \emptyset$$
$$= \{K\} \text{ otherwise.}$$

We need to prove (1) γ is well defined, (2) γ is a selection function and (3) $K * \alpha = (\bigcap \gamma(K\bot\neg\alpha)) + \alpha$.

1. If $\alpha \notin K$ then $K\bot\alpha = K\bot\beta = \{K\}$, hence, $\alpha \in K$ If $\gamma(K\bot\alpha) = \gamma(K\bot\beta) = \{K\}$. and $K \bot \alpha = K \bot \beta$ then, by Lemma 4.7, $Cn(\alpha) = Cn(\beta)$ and, by *extensionality* $K - \alpha = K - \beta$. It follows that $\gamma(K \bot \alpha) = \gamma(K \bot \beta)$.

2. It follows directly from construction that $\gamma(K\bot\alpha) \subseteq K$ Now if $K\bot\alpha \neq \emptyset$ then, by Lemma 4.4, $\alpha \notin Cn(\emptyset)$. It follows that $\neg\alpha$ is consistent and, by *consistency*, $K*\neg\alpha$ is consistent. Since by *success* $\neg\alpha \in K*\neg\alpha$, we have that $\alpha \notin K*\neg\alpha$. By *upper bound property* (Lemma 4.3), there is X such that $K*\neg\alpha \cap K \subseteq X \in K\bot\alpha$. Hence $\gamma(K\bot\alpha) \neq \emptyset$

3. By construction $K \cap K * \alpha \subseteq \bigcap \gamma(K\bot\neg\alpha)$. It follows that $(K \cap K * \alpha) + \alpha \subseteq (\bigcap \gamma(K\bot\neg\alpha)) + \alpha$. By *distributivity* $(K \cap K*\alpha) + \alpha = (K + \alpha) \cap ((K*\alpha) + \alpha)$. By *success* and *inclusion* $(K + \alpha) \cap ((K * \alpha) + \alpha) = K * \alpha$. We conclude that $K * \alpha \subseteq (\bigcap \gamma(K\bot\neg\alpha)) + \alpha$.

Now suppose that $\beta \notin K*\alpha$.

If $\neg\alpha \notin K$ then $K + \alpha$ is consistent and, by *vacuity*, $K*\alpha = K+\alpha$. Furthermore, $K\bot\neg\alpha = \{K\}$. Hence, $(\bigcap \gamma(K\bot\neg\alpha)) + \alpha = K * \alpha = K + \alpha$.

So let $\neg\alpha \in K$ and suppose that $\neg\alpha \in Cn((K \cap K*\alpha) \cup \{(\alpha \to \beta) \to \neg\alpha\}$. By *deduction* and *closure* $((\alpha \to \beta) \to \neg\alpha) \to \neg\alpha \in K \cap K*\alpha$. By *supraclassicality* $\alpha \to \beta \in K \cap K * \alpha$. By *success* $\alpha \in K * \alpha$, hence, $\beta \in K * \alpha$ which contradicts the hypothesis.

It follows that $\neg\alpha \notin Cn((K \cap K * \alpha) \cup \{(\alpha \to \beta) \to \neg\alpha\})$.

Since $\neg\alpha \in K$ and $(\alpha \to \beta) \to \neg\alpha \in C_{CPL}(\neg\alpha)$, we have that $(\alpha \to \beta) \to \neg\alpha \in K$. By *upper bound property* (Lemma 4.3), there is X such that $(K \cap K * \alpha) \cup \{(\alpha \to \beta) \to \neg\alpha\} \subseteq X \in K\bot\neg\alpha$. Of course $\alpha \to \beta \notin X$, otherwise $\neg\alpha \in X$ which would contradict the definition of X. Hence $\alpha \to \beta \notin \bigcap \gamma(K\bot\neg\alpha)$ and $\beta \notin (\bigcap \gamma(K\bot\neg\alpha)) + \alpha$. We conclude that $(\bigcap \gamma(K\bot\neg\alpha)) + \alpha \subseteq K * \alpha$. □

Theorems 4.10 and 4.11 showed that given a withdrawal operation an AGM revision can be defined via the Levi identity and given an AGM revision an AGM contraction can be defined via Harper identity. The following theorems show that the relation between contraction and revision is even stronger.

Theorem 4.13 [Mak87] *Let $-$ be an AGM contraction over K and let $*$ be a revision over K obtained via Levi identity from $-$. For every sentence $\alpha \in \mathscr{L}$ we have that $K \cap (K * \neg\alpha) = K - \alpha$.*

Proof If $\alpha \notin K$ then $K + \neg\alpha$ is consistent and by *vacuity of revision* $K * \neg\alpha = K + \neg\alpha$. It follows that $K \cap K * \neg\alpha = K \cap (K + \neg\alpha) = K$. By *vacuity of contraction* $K - \alpha = K$. Hence, $K \cap K * \neg\alpha = K - \alpha$.

If $\alpha \in K$ then by *recovery and inclusion of contraction* $K = (K - \alpha) + \alpha$. It follows that:

$$
\begin{aligned}
K \cap K * \neg\alpha &= K \cap (K - \alpha + \neg\alpha) \\
&= (K - \alpha + \alpha) \cap (K - \alpha + \neg\alpha) \\
&= K - \alpha + (Cn(\alpha) \cap Cn(\neg\alpha)) \quad \text{by distributivity} \\
&= K - \alpha + Cn(\emptyset) \\
&= K - \alpha \qquad\qquad\qquad\qquad\qquad\qquad\quad \square
\end{aligned}
$$

Theorem 4.14 [Mak87] *Let $*$ be an AGM revision over K and let $-$ be a contraction over K obtained via Harper identity from $*$. For every sentence $\alpha \in \mathscr{L}$ we have that $(K - \neg\alpha) + \alpha = K * \alpha$.*

Proof If $\neg\alpha \notin K$ then $K - \neg\alpha + \alpha = K + \alpha$ by *vacuity of contraction*. In this case, since $K + \alpha$ consistent, $K * \alpha = K + \alpha$ by *vacuity of revision*. Hence, $(K - \neg\alpha) + \alpha = K * \alpha$

If $\neg\alpha \in K$ then:

$$
\begin{aligned}
(K - \neg\alpha) + \alpha &= (K \cap K * \alpha) + \alpha \\
&= (K + \alpha) \cap (K * \alpha + \alpha) \quad \text{by distributivity} \\
&= (K + \alpha) \cap (K * \alpha) \qquad \text{by success of revision} \\
&= K * \alpha \qquad \text{by inclusion of revision} \qquad\qquad \square
\end{aligned}
$$

4.3 Belief Base Theory

Belief base theory studies the dynamics of epistemic states represented as arbitrary sets of sentences B. It studies a belief system that admits four types of epistemic attitudes with respect to a sentence α :

reject: α is not consistent with B.
explicitly accepted: $\alpha \in B$.

implicitly accepted: $\alpha \in Cn(B)$, but $\alpha \notin B$.

undetermined: α consistent with B and $\alpha \notin Cn(\alpha)$.

If $\alpha \in Cn(B)$ we will simply say that α is accepted, i.e., α is accepted if it is implicitly or explicitly accepted.

Belief base theory admits the same three types of belief change of AGM theory: expansion, revision, and contraction. *Expansion* turns a sentence α to be accepted and is defined simply as $B + \alpha = B \cup \{\alpha\}$. Like AGM theory *contraction* and *revision* are defined through a set of rationality postulates.

Some authors such as Nebel [Neb90] use the term belief base in reference to a finite representation of a belief set. Our approach differs from Nebel and follows authors such as Fuhrmann, Hansson, and Wassermann [Fuh97, Han99, Was00]. For this second group of authors belief bases are a distinct belief system. In this belief system it is possible to distinguish what an agent explicitly believes from what is merely a consequence of its explicit beliefs.

The following example illustrates the difference between belief set approach and belief base approach:

Example. 1.15 [Han99] I believe that Paris is the capital of France (α). I also believe that there is milk in the fridge (β). Hence, I believe that Paris is the capital of France if and only if there is milk in the fridge ($\alpha \leftrightarrow \beta$). I open my fridge and I see that there is no milk in it ($\neg\beta$). I cannot keep both my beliefs in α and in $\alpha \leftrightarrow \beta$ at the same time.

Belief set approach: Both α and $\alpha \leftrightarrow \beta$ are elements from the belief set. After noticing that there is no milk in the fridge I have to chose between retaining α or $\alpha \leftrightarrow \beta$ in my belief set. The removal of $\alpha \leftrightarrow \beta$ does not follow automatically, it must be guaranteed by a selection mechanism.

Belief base approach: The sentence $\alpha \leftrightarrow \beta$ is merely derived from the others. When β is removed $\alpha \leftrightarrow \beta$ is automatically removed.

Both approaches have strong advantages. In belief set approach equivalent epistemic states are treated equivalently, i.e., this approach abstracts the syntactic form of the beliefs. Belief base approach, on the other hand, is more expressive and more interesting from a computational point of view. From a philosophical point of view there are authors, such as Harman [Har86] , who defend the distinction between explicit beliefs and implicit belief. Other authors, such as Stalnaker [Sta84] , defend the opposite.

4.3.1 Base Contraction

Belief base contraction is an operation in a belief base B that makes a sentence α to become undetermined. Typically an agent accepts (implicitly or explicitly) a sentence α and after the contraction it should have no opinion about α. Furthermore, it is desirable that the change in the agent's belief base is minimal.

Most AGM postulates for contraction should also hold in belief bases:

(**success**) If $\alpha \notin Cn(\emptyset)$ then $\alpha \notin Cn(B - \alpha)$
(**inclusion**) $B - \alpha \subseteq B$
(**vacuity**) If $\alpha \notin Cn(\beta)$ then $B - \alpha = B$
(**extensionality**) If $Cn(\alpha) = Cn(\beta)$ then $B - \alpha = B - \beta$

Since epistemic states are now represented as arbitrary sets of sentence, it is obvious that closure should not hold. For this reason success was adapted. Furthermore, inclusion now is much more controversial (see [RW08]).

Example 4.16 Consider an agent whose epistemic state is represented by the following base in the Description Logics $\mathcal{ALC}$:

$$B = \{Man \sqsubseteq Mortal$$
$$Man(SOCRATES)\}$$

This agent must accept the sentence $Mortal(SOCRATES)$. It may want to remove this sentence from its belief. In the belief set approach one of its possibilities is to believe that every man *except* Socrates is mortal. In belief base approach, however, since it must respect inclusion the only possibilities the agent have are: B, $\{Man \sqsubseteq Mortal\}$, $\{Man(SOCRATES)\}$ and $\emptyset$. Of course the third is not possible by success.

Although controversial, inclusion is mostly accepted in the belief base literature. Recovery, however, is even more problematic. It is not even compatible with inclusion and success:

Example 4.17 [Han99] Let $-$ be an operation over $B = \{p \wedge q\}$ that satisfies success and inclusion. By these two postulates we have that $B - p \subseteq \emptyset$ i.e. $B = \emptyset$. However, $\emptyset + p = \emptyset \cup \{p\} = \{p\} \neq B$. Hence, in this case if $-$ satisfies success and inclusion then it cannot satisfy recovery.

The above example is a strong justification for not considering recovery in belief bases. However, as argued in Sect. 4.2.1, recovery postulate is an important minimality postulate. Hence, in order to drop this postulate we need a new postulate to replace it. In the literature we find two postulates suggested for this purpose [Han91]:

(**relevance**) If $\beta \in B$ and $\beta \notin B - \alpha$ then there is a B' such that $B - \alpha \subseteq B' \subseteq B$ and $\alpha \notin Cn(B')$, but $\alpha \in Cn(B' \cup \{\beta\})$.

(**core-retainment**) If $\beta \in B$ and $\beta \notin B - \alpha$ then there is a B' such that $B' \subseteq B$ and $\alpha \notin Cn(B')$, but $\alpha \in Cn(B' \cup \{\beta\})$.

The main idea behind these postulates is to guaranty that a sentence β may be removed only if it somehow helps to derive α. Notice that relevance is stronger than core-retainment i.e. every operation that satisfies relevance satisfies core-retainment, but the opposite is not always true.

Furthermore, vacuity become redundant in the presence of core-retainment (or relevance which is stronger) and inclusion:

Proposition 1.18 [Han99] *Let – be an operation over B that satisfies core-retainment and inclusion then – satisfies vacuity.*

Proof Let $\alpha \notin Cn(B)$ and suppose that $B - \alpha \neq B$. By *inclusion* $B - \alpha \subset B$ and, hence, there is $\beta \in B$ such that $\beta \notin B - \alpha$. By *core-retainment* there is $B' \subseteq B$ such that $\alpha \notin Cn(B')$, but $\alpha \in Cn(B' \cup \{\beta\})$. Since $\alpha \notin Cn(B)$, by *monotonicity*, $\alpha \notin Cn(K' \cup \{\beta\})$ which is a contradiction. We conclude that $K - \alpha = K$

In AGM theory extensionality guarantees that equivalent sentences contracted from the same belief set result in the same belief set. In belief bases extensionality is too weak for this purpose. The following postulate is used instead:

(**uniformity**) If for every subset B' of B it holds that $\alpha \in Cn(B')$ iff $\beta \in Cn(B')$ then it holds that $B - \alpha = B - \beta$.

In the presence of uniformity and core-retainment both vacuity and extensionality are redundant. It follows that we need only four postulates to characterize belief base contraction: success, inclusion, uniformity, and a minimality criterion that may be relevance or core-retainment. Each possible minimality criterion corresponds to one of the constructions presented in the following sections: the former corresponds to partial meet contraction and the latter to kernel contraction.

Partial Meet Base Contraction

Partial meet contraction in belief bases is very similar to partial meet contraction in belief sets. The only difference here is that it starts with an arbitrary set of sentences B instead of a belief set K. The rest is identical: first get the remainder set $B \perp \alpha$ and then, given a selection function γ, get the intersection of the elements chosen by γ :

$$B -_\gamma \alpha = \bigcap \gamma(B \perp \alpha)$$

This construction is completely characterized by success, inclusion, uniformity and relevance. Formally:

Representation Theorem 4.19 [HW02] *Let $\langle \mathscr{L}, Cn \rangle$ be a monotonic and compact logic. The operation – is a partial meet contraction for a belief base B iff – satisfies inclusion, success, uniformity, and relevance.*

(**construction $\Rightarrow$ postulates**)

success: If $\alpha \notin Cn(\emptyset)$ then by Lemma 4.4 $B \perp \alpha \notin \emptyset$. $\alpha \notin Cn(X)$ for $X \in B \perp \alpha$, hence, $\alpha \notin Cn(\bigcap \gamma(B \perp \alpha))$

inclusion: Follows directly from the construction.

uniformity: Follows from Lemma 4.6.

relevance: If $\alpha \in Cn(\emptyset)$ then $\bigcap \gamma(B \perp \alpha) = B$ and relevance is vacuously satisfied.

Let $\alpha \notin Cn(\emptyset)$ and $\beta \in B \setminus B - \alpha$. Then there is some $X \in \gamma(B \perp \alpha)$ such that $\beta \notin X$. Of course $B - \alpha \subseteq X \subseteq B$ and $\alpha \notin Cn(X)$. Furthermore, since X is maximal and $\beta \notin X$ then $\alpha \in Cn(X \cup \{\beta\})$.

(**postulates $\Rightarrow$ construction**)

Let – be an operation over B that satisfies success, inclusion, uniformity and relevance and let γ be a selection function defined as follows:

$$\gamma(B\perp\alpha) = \{X \in B\perp\alpha : B - \alpha \subseteq X\} \text{ if } B\perp A \neq \emptyset$$
$$= \{B\} \text{ otherwise.}$$

By uniformity and Lemma 4.6 we have that γ is well defined. We will prove (1) if $B\perp\alpha \neq \emptyset$ then $\gamma(B\perp\alpha) \neq \emptyset$ and (2) $\bigcap \gamma(B\perp\alpha) = B - \alpha$.

1. If $B\perp\alpha \neq \emptyset$ then, by Lemma 4.4, $\alpha \notin Cn(\emptyset)$. By *success* and *inclusion* $\alpha \notin B - \alpha \subseteq B$. It follows, by *upper bound property* (Lemma 4.3), that there is X such that $B - \alpha \subseteq X \in B\perp\alpha$ and, hence, $\gamma(B\perp\alpha) \neq \emptyset$.
2. If $B\perp\alpha = \emptyset$ then, by Lemma 4.4, $\alpha \in Cn(\emptyset)$. It follows that $\bigcap \gamma(B\perp\alpha) = B$. By *relevance*, if $\beta \in B$ and $\beta \notin B - \alpha$ then there is B' such that $\alpha \notin Cn(B')$. Since $\alpha \in Cn(\emptyset)$, there is no such B', hence, $B \subseteq B - \alpha$. It follows, by *inclusion*, that $B - \alpha = B$. Let $B\perp\alpha \neq \emptyset$. It follows, by construction, that $B - \alpha \subseteq \bigcap \gamma(B\perp\alpha)$. Now let $\beta \notin B - \alpha$. If $\beta \notin B$ then trivially $\beta \notin \bigcap \gamma(B\perp\alpha)$. So let $\beta \in B$. By *relevance* there is B' such that $B - \alpha \subseteq B' \subseteq B$, $\alpha \notin Cn(B')$ and $\alpha \in Cn(B' \cup \{\beta\})$. By *upper bound property* (Lemma 4.3) there is X such that $B' \subseteq X \in B\perp\alpha$. Since $B' \subseteq X$ $\alpha \in Cn(B' \cup \{\beta\})$ and $\alpha \notin Cn(X)$, we have that $\beta \notin X$. It follows that $\beta \notin X \in \gamma(B\perp\alpha)$, hence, $\beta \notin \bigcap \gamma(B\perp\alpha)$.   □

Notice that this Representation Theorem holds for every compact while Representation Theorem 4.9 holds only for logics that satisfies the AGM assumptions.

Kernel Base Contraction

In partial meet contraction maximal subsets of B that do not imply α are considered. Kernel contraction [Han94], instead of considering these maximal subsets of B, considers minimal subsets of B that do imply α. Formally:

Definition 4.20 (*kernel*) [Han94] Let B be a belief base, i.e., $B \subseteq \mathcal{L}$ and let $\alpha \in \mathcal{L}$. The set $B \perp\!\!\!\perp \alpha$ is a set such that $X \in B \perp\!\!\!\perp \alpha$ iff:

1. $X \subseteq B$ (X *is a subset of* B).
2. $\alpha \in Cn(X)$ (*that implies* α)
3. if $X' \subset X$ *then* $\alpha \notin Cn(X')$ (*and is minimal*)

Lemma 4.21 *Let B be a set of sentences in a compact logic $\langle \mathcal{L}, Cn \rangle$ and let B' be a subset of B and β be an element of B such that $\alpha \in Cn(B' \cup \{\beta\})$, but $\alpha \notin Cn(B')$. There is a set $X \subseteq B' \cup \{\beta\}$ such that $\beta \in X \in B \perp\!\!\!\perp \alpha$*

Proof By *compactness* there is a finite subset B'' of $B' \cup \{\beta\}$ such that $\alpha \in Cn(B'')$. Suppose without loss of generality that $B'' = \{\beta_0, \beta_1, \ldots, \beta_n\}$ and that $\beta_0 = \beta$. Let $X_0 = B''$ and for every i such that $1 \leq i \leq n$ let X_i be defined recursively as follows:

$$X_i = \begin{cases} X_{i-1} \setminus \{\beta_i\} & \text{if } \alpha \in Cn(X_{i-1} \setminus \{\beta_i\}) \\ X_{i-1} & \text{otherwise.} \end{cases}$$

Notice that $X_n \subseteq B'' \subseteq B' \cup \{\beta\} \subseteq B$ $\beta \in X_n$ and $\alpha \in Cn(X_n)$. Furthermore, if $X \subset X_n$ there is $\beta_i \in X_n \setminus X$. By construction, either $\alpha \notin Cn(X_{i-1} \setminus \{\beta_i\})$ or

$\beta_i = \beta_0 = \beta$. In the former case $\alpha \notin Cn(X) \subseteq Cn(X_i)$. In the later case, since $\alpha \notin Cn(B')$ then $\alpha \notin Cn(X_n \setminus \{\beta\})$ and, hence, $\alpha \notin Cn(X)$. □

Lemma 4.22 *Let B be a set of sentence and α and β be sentences in a logic $\langle \mathcal{L}, Cn \rangle$. $B \perp\!\!\!\perp \alpha = B \perp\!\!\!\perp \beta$ if and only if for every subset B' of B it holds that $\alpha \in Cn(B')$ iff $\beta \in Cn(B')$*

Proof ($\Rightarrow$) Let $B' \subseteq B$ and $\alpha \in Cn(B')$. Then by the previous lemma there is $X \subseteq B'$ such that $X \in B \perp\!\!\!\perp \alpha$. If $\beta \notin Cn(B')$ then $\beta \notin Cn(X)$ and, hence, $X \notin B \perp\!\!\!\perp \beta$.

($\Leftarrow$) This holds trivially. □

Each element of the kernel $B \perp\!\!\!\perp \alpha$ is called an α-kernel. In kernel contraction, at least one element of each α-kernel chosen by a function called incision function is removed. Incision function plays a role in kernel contraction analogous to the role played by selection function in partial meet contraction.

Definition 4.23 (*incision function*) [Han94] Let B be a belief base. An incision function for B is a function σ such that for every α :

1. $\sigma(B \perp\!\!\!\perp \alpha) \subseteq \bigcup B \perp\!\!\!\perp \alpha$.
2. if $\emptyset \neq X \in B \perp\!\!\!\perp \alpha$ then $X \cap \sigma(B \perp\!\!\!\perp \alpha) \neq \emptyset$.

Intuitively incision function chooses the sentences which are less entrenched in the agents beliefs.

Let σ be an incision function for a belief base B. The kernel contraction operation $-_\sigma$ is formally defined as:

$$B -_\sigma \alpha = B \setminus \sigma(B \perp\!\!\!\perp \alpha)$$

Kernel contraction is completely characterized by the following postulates: success, inclusion, uniformity, and core-retainment. Formally:

Representation Theorem 4.24 [HW02] *Let $\langle \mathcal{L}, Cn \rangle$ be a monotonic and compact logic. The operation $-$ is a kernel contraction for a belief base B iff $-$ satisfies inclusion, success, uniformity, and core-retainment.*

Proof (**construction $\Rightarrow$ postulates**)

success: Suppose that $\alpha \in B -_\sigma \alpha$ for some $\alpha \notin Cn(\emptyset)$. By Lemma 4.21 there is $X \in B \perp\!\!\!\perp \alpha$ such that $X \subseteq B -_\sigma \alpha$. Since $\alpha \notin Cn(\emptyset)$, we have that $X \neq \emptyset$ and, by definition, $\sigma(B \perp\!\!\!\perp \alpha) \cap X \neq \emptyset$. It follows that $X \not\subseteq B -_\sigma \alpha$ which is a contraction.
inclusion: Follows directly from construction.
uniformity: Follows directly from Lemma 4.22.
core-retainment: Let $\beta \in B$ and $\beta \notin B -_\sigma \alpha$. It follows that $\beta \in \sigma(B \perp\!\!\!\perp \alpha)$ and, hence, $\beta \in X \in B \perp\!\!\!\perp \alpha$. Of course $X \setminus \{\beta\} \subseteq B$ and $\alpha \in Cn(X) = Cn((X \setminus \{\beta\}) \cup \{\beta\})$. Furthermore, since X is minimal, $\alpha \notin Cn(X \setminus \{\beta\})$
(**postulates $\Rightarrow$ construction**)

Let $-$ be an operation over B that satisfies success, inclusion, uniformity and core-retainment and let σ be a selection function defined as follows:

$$\sigma(B \perp\!\!\!\perp \alpha) = B \setminus (B - \alpha)$$

By Lemma 4.22, σ is well defined. We will prove that (1) $B - \alpha = B -_{\sigma} \alpha$ and (2) σ is an incision function (i.e., $\sigma(B \perp\!\!\!\perp \alpha) \subseteq \bigcup B \perp\!\!\!\perp \alpha$ and if $\emptyset \neq X \in B \perp\!\!\!\perp \alpha$ then $X \cap \sigma(B \perp\!\!\!\perp \alpha) \neq \emptyset$).

1. Follows trivially by *inclusion*.
2. If $\beta \in B \setminus B - \alpha$ then, by *core-retainment*, there is $B' \subseteq B$ such that $\alpha \notin Cn(B')$, but $\alpha \in Cn(B' \cup \{\beta\})$. By Lemma 4.21, there is $X \in B \perp\!\!\!\perp \alpha$ such that $X \subseteq B' \cup \{\beta\}$ and $\beta \in X$. Hence, $\beta \in \bigcup B \perp\!\!\!\perp \alpha$.

Now let $\emptyset \neq X \in B \perp\!\!\!\perp \alpha$ and suppose that $X \cap \sigma(B \perp\!\!\!\perp \alpha) = \emptyset$. It follows that $\alpha \notin Cn(\emptyset)$ and, by (1), $B - \alpha = B \setminus \sigma(B \perp\!\!\!\perp \alpha)$. It follows that $X \subseteq B - \alpha$, but then $\alpha \in B - \alpha$ which contradicts the *success*, hence, $X \cap \sigma(B \perp\!\!\!\perp \alpha) \neq \emptyset$.   □

4.3.2 Belief Base Revision

Belief base revision is an operation in a belief base B that makes a sentence α to become explicitly accepted by the agent in a consistent way. Typically an agent rejects a sentence α (i.e. α is inconsistent with the beliefs of the agent) and after the revision the agent accepts α and still has a consistent belief base. This operation should also change the agent's belief base minimally.

In AGM theory it was shown that it is possible to define a revision from an AGM contraction via Levi identity:

$$B * \alpha = (B - \neg\alpha) + \alpha$$

In the case of belief base there is yet another possibility of defining revision using a reverted version of Levi identity [Han93] :

$$B * \alpha = (B + \alpha) - \neg\alpha$$

In AGM theory the second possibility is not allowed since $K + \alpha$ is typically trivial. This is not the case in belief bases, since it is possible in belief bases to have distinct inconsistent bases. For example:

$$B_1 = \{p \wedge \neg p\}$$
$$B_2 = \{q, p, \neg p\}$$

Belief base revision constructed via Levi identity is called *internal revision* while the revision constructed via reversed Levi identity is called *external revision*. Both internal and external revisions satisfy the following postulates:

(**success**) $\alpha \in B * \alpha$

(**inclusion**) $B * \alpha \subseteq B + \alpha$

(**non-contravention**) If $\neg\alpha \notin Cn(\emptyset)$ then $\neg\alpha \notin Cn(B * \alpha)$.

Besides these postulates both revisions must satisfy some minimally criterion, core-retainment or relevance, depending whether the contraction used in the construction is kernel or partial meet.

(**core-retainment**) If $\beta \in B$ and $B \notin B * \alpha$ then there is a B' such that $B' \subseteq B \cup \{\alpha\}$, $\neg\alpha \notin Cn(B')$ and $\neg\alpha \in Cn(B' \cup \{\beta\})$.

(**relevance**) If $\beta \in B$ and $B \notin B * \alpha$ then there is a B' such that $B - \alpha \subseteq B' \subseteq B \cup \{\alpha\}$, $\neg\alpha \notin Cn(B')$ and $\neg\alpha \in Cn(B' \cup \{\alpha\})$.

Furthermore, internal revision satisfies uniformity:

(**uniformity**) If it holds that for every $B' \subseteq B$, $\neg\alpha \in Cn(B')$ iff $\neg\beta \in Cn(B')$ then $B \cap (B * \alpha) = B \cap (B * \beta)$.

Besides the success, inclusion, non-contravention and one of the minimality postulates, external revision satisfies a weak version of uniformity and a postulate called pre-expansion.

(**weak uniformity**) If $\alpha, \beta \in B$ and it holds that for every $B' \subseteq B$ $\neg\alpha \in Cn(B')$ iff $\neg\beta \in Cn(B')$ then $B \cap (B * \alpha) = B \cap (B * \beta)$.

(**pre-expansion**) $(B + \alpha) * \alpha = B * \alpha$

Summing up, four types of belief base revision were presented:

internal kernel revision: $B *_\sigma \alpha = (B \setminus \sigma(B \perp\!\!\!\perp \neg\alpha)) + \alpha$

external kernel revision: $B *_\sigma \alpha = (B + \alpha) \setminus \sigma((B + \alpha) \perp\!\!\!\perp \neg\alpha)$

internal partial meet revision: $B *_\gamma \alpha = (\bigcap \gamma(B \perp \neg\alpha)) + \alpha$

external partial meet revision: $B *_\gamma \alpha = \bigcap \gamma((B + \alpha) \perp \neg\alpha)$

Each of these constructions is characterized by a set of postulates. The following representation theorems proves these characterization. Each of these theorems assume that the underlying logic satisfies monotonicity, compactness and α-local noncontravention.

Representation Theorem 4.25 (IKR) [HW02] *Let $\langle \mathscr{L}, Cn \rangle$ be a logic that satisfies monotonicity, compactness, and α -local non- contravention. A operation $*$ over a belief base B is an internal kernel revision iff it satisfies: success, inclusion, non-contravention, uniformity, and core-retainment.*

Proof (**construction $\Rightarrow$ postulates**)

success and inclusion: Follows directly from construction.

uniformity: Follows from Lemma 4.22.

non-contravention: If $\neg\alpha \notin Cn(\emptyset)$ then, by the same argument of previous Representation Theorem, $\neg\alpha \notin Cn(B \setminus \sigma(B \perp\!\!\!\perp \neg\alpha))$. It follows, by α-*local non-contravention* that $\neg\alpha \notin Cn(B *_\sigma \alpha)$.

core-retainment: If $\neg\alpha \in Cn(\emptyset)$ then $B \perp\!\!\!\perp \neg\alpha = \emptyset$. It follows that $B \subseteq B *_\sigma \alpha$ and, hence, core-retainment follows vacuously.

Let $\neg\alpha \notin Cn(\emptyset)$ and $\beta \in B \setminus B *_\sigma \alpha$. By Lemma 4.21, there is X such that $\beta \in X \subseteq B \perp\!\!\!\perp \neg\alpha$. Of course $X \setminus \{\beta\} \subseteq B$ and $\alpha \in Cn(X) = Cn((X \setminus \{\beta\}) \cup \{\beta\})$. Furthermore, since X is minimal and $\beta \in X$ then $\alpha \notin Cn(X \setminus \{\beta\})$.

(postulates ⇒ construction)

Let – be an operation over B that satisfies success, inclusion, uniformity, core-retainment, and non-contravention. Let σ be an incision function defined as follows:

$$\sigma(B \perp\!\!\!\perp \neg\alpha) = B \setminus (B \cap B * \alpha)$$

By Lemma 4.22 and *uniformity*, σ is well defined. It will be proved that (1) σ is an incision function and (2) $B * \alpha = B *_\sigma \alpha$.

1. Let $\beta \in \sigma(B \perp\!\!\!\perp \neg\alpha)$ then $\beta \in B \setminus (B \cap B * \alpha) = B \setminus B * \alpha$. It follows, by *core-retainment*, that there is a $B' \subseteq B$ such that $\neg\alpha \notin Cn(B')$ and $\neg\alpha \in Cn(B' \cup \{\beta\})$. By Lemma 4.21, there is X such that $\beta \in X \in B \perp\!\!\!\perp \neg\alpha$. Hence, $\beta \in \bigcup B \perp\!\!\!\perp \neg\alpha$.

Now let $\emptyset \neq X \in B \perp\!\!\!\perp \neg\alpha$. It follows that $\neg\alpha \notin Cn(\emptyset)$ and $\neg\alpha \in Cn(X)$. If $X \subseteq B \perp\!\!\!\perp \neg\alpha$ then $\neg\alpha \in B * \alpha$ contracting *non-contravention*. Hence, $X \not\subseteq B \perp\!\!\!\perp \neg\alpha$. It follows that there is $\beta \in X$ such that $\beta \in B \setminus (B \cap B * \alpha) = \sigma(B \perp\!\!\!\perp \neg\alpha)$. We conclude that $X \cap \sigma(B \perp\!\!\!\perp \neg\alpha) \neq \emptyset$.

$$\begin{aligned}
B *_\sigma \alpha &= (B \setminus \sigma(B \perp\!\!\!\perp \neg\alpha)) + \alpha \\
&= (B \setminus (B \setminus (B * \alpha))) + \alpha \text{ by inclusion} \\
&= B * \alpha + \alpha \\
&= B * \alpha \qquad\qquad\qquad \text{by success}
\end{aligned}$$

$\square$

Representation Thoerem 4.26 (EKR) [HW02] *Let $\langle \mathscr{L}, Cn \rangle$ be a logic that satisfies monotonicity, compactness, and α-local non-contravention. A operation $*$ over a belief base B is a external kernel revision iff it satisfies: success, inclusion, non-contravention, weak uniformity, pre-expansion, and core-retainment.*

Proof (**construction ⇒ postulates**)

inclusion and pre-expansion: Follows directly from construction.

non-contravention: Very similar with success in Representation Theorem 4.24.

weak uniformity: Follows from Lemma 4.22.

core-retainment: If $B \setminus B *_\sigma \alpha$ then $\beta \in \sigma(B + \alpha \perp\!\!\!\perp \neg\alpha)$. It follows that there is X such that $\beta \in X \in B + \alpha \perp\!\!\!\perp \neg\alpha$. Notice that $X \setminus \{\beta\} \subseteq B + \alpha$, $\neg\alpha \notin Cn(X \setminus \{\beta\})$ and $\neg\alpha \in Cn(X)$.

success: If $X \in B + \alpha \perp\!\!\!\perp \neg\alpha$ then $\alpha \notin X$ by α-*local non contravention*. It follows that $\alpha \notin \sigma(B + \alpha \perp\!\!\!\perp \neg\alpha)$. Hence, $\alpha \in B *_\sigma \alpha$.

(**postulates ⇒ construction**) Let – be an operation over B that satisfies success, inclusion, non-contravention, weak uniformity, pre-expansion, and core-retainment. Let σ be an incision function defined as follows:

$$\sigma(B \perp\!\!\!\perp \neg\alpha) = B \setminus B * \alpha$$

By Lemma 4.22 and weak uniformity, σ is well defined. Furthermore, the proof that σ is an incision function is very similar to the one presented in previous Representation Theorem. The only thing left to show is that $B * \alpha = B *_\sigma \alpha$.

$$B *_\sigma \alpha = (B + \alpha) \setminus \sigma(B + \alpha \perp\!\!\!\perp \neg\alpha)$$
$$= (B + \alpha) \setminus ((B + \alpha) \setminus ((B + \alpha) * \alpha))$$
$$= (B + \alpha) * \alpha \qquad \text{by inclusion}$$
$$= B * \alpha \qquad \text{by success}$$

□

Representation Theorem 4.27 (IPMR) [HW02] *Let $\langle \mathcal{L}, Cn \rangle$ be a logic that satisfies monotonicity, compactness, and α-local non contravention. A operation $*$ over a belief base B is a internal partial meet revision iff it satisfies: success, inclusion, non-contravention, uniformity and relevance.*

Proof (**construction $\Rightarrow$ postulates**)

inclusion and success: Follows directly from construction.
uniformity: Follows from Lemma 4.6.
non-contravention: If $B \perp \neg\alpha = \emptyset$ then, by Lemma 4.4, $\neg\alpha \in Cn(\emptyset)$ and the postulate is vacuously satisfied.
If $B \perp \neg\alpha \neq \emptyset$ then $\neg\alpha \notin Cn(\bigcap \gamma(B \perp \neg\alpha))$ and, by α-*local non contravention* $\neg\alpha \notin Cn(B *_\gamma \alpha)$.
relevance: If $\beta \in B \setminus B *_\gamma \alpha$ then there is some X such that $\beta \notin X \in \gamma(B \perp \neg\alpha)$. $B *_\gamma \alpha \subseteq X \cup \{\alpha\} \subseteq B + \alpha \neg\alpha \notin Cn(X \cup \{\alpha\})$ by α-*local non contravention*. Furthermore, $\neg\alpha \in Cn(X \cup \{\beta\})$ and, by *monotonicity*, $\neg\alpha \in Cn(X \cup \{\alpha, \beta\})$.

(**postulates $\Rightarrow$ construction**) Let $-$ be an operation over B that satisfies success, inclusion, non-contravention, uniformity, and relevance. Let γ be a selection function defined as follows:

$$\gamma(B \perp \alpha) = \{X \in B \perp \alpha : B \cap B * \neg\alpha \subseteq X\} \text{ if } B \perp \alpha \neq \emptyset$$
$$= \{B\} \text{ otherwise.}$$

From *uniformity* and Lemma 4.6, we have that γ is well defined. It will be proved that (1) if $B \perp \alpha \neq \emptyset$ then $\gamma(B \perp \alpha) \neq \emptyset$ and (2) $B * \alpha = B *_\gamma \alpha$.

1. If $B \perp \alpha \neq \emptyset$ then, by Lemma 4.4, $\alpha \notin Cn(\emptyset)$. By *non-contravention* $\alpha \notin B * \neg\alpha$. It follows, by *upper bound property* (Lemma 4.3) that there is X such that $B \cap B * \neg\alpha \subseteq X \in B \perp \alpha$. Hence, $\gamma(B \perp \alpha) \neq \emptyset$.
2. By *inclusion* $B * \alpha \subseteq B + \alpha \subseteq (B \cap B * \alpha) + \alpha$. It follows that $B * \alpha \subseteq \bigcap \gamma(B \perp \neg\alpha) + \alpha$.

Now let $\beta \notin B * \alpha$. If $\beta \notin B$ then $\beta \notin B *_\gamma \alpha$ trivially. So let $\beta \in B$. By *relevance*, there is B' such that $B * \alpha B' \subseteq B + \alpha \neg\alpha \notin Cn(B')$ and $\neg\alpha \in Cn(B' \cup \{\beta\})$. By *success*, $\alpha \in B'$ and, since $\neg\alpha \notin Cn(B')$, by α-*local non contravention*, $\neg\alpha \notin Cn(B' \setminus \{\alpha\})$. It follows, by *upper bound property* (Lemma 4.3) that there is X such that $B' \setminus \{\alpha\} \subseteq X \in B \perp \neg\alpha$. Since $B \cap B * \alpha \subseteq B \setminus \{\alpha\} \subseteq X$, we have that $X \in \gamma(B \perp \neg\alpha)$. Since $\neg\alpha \in Cn(B' \cup \{\beta\})$, by α-*local non contravention*, we have that $\neg\alpha \in Cn((B' \setminus \{\alpha\}) \cup \{\beta\})$. Hence, by *monotonicity* $\neg\alpha \in Cn(X \cup \{\beta\})$. Since $\beta \notin X$, we have that $\beta \notin \bigcap \gamma(B \perp \neg\alpha)$. Hence, $B *_\gamma \alpha \subseteq B * \alpha$. □

Representation Theorem 4.28 (EPMR) [HW02] *Let $\langle \mathcal{L}, Cn \rangle$ be a logic that satisfies monotonicity, compactness and α-local non-contravention. A operation $*$ over a belief base B is a external partial meet revision iff it satisfies: success, inclusion, non-contravention, weak uniformity, pre-expansion and relevance.*

Proof (**construction** $\Rightarrow$ **postulates**)

inclusion, pre-expansion and non-contravention: Follows directly from construction.

weak uniformity: Follows from Lemma 4.6.

success: If $B + \alpha \perp \neg\alpha = \emptyset$ then it follows trivially. If $X \in B + \alpha \perp \neg\alpha$ then, by *alpha-local non contravention*, $\alpha \in X$. Hence, $\alpha \in B *_\gamma \alpha$.

(**postulates** $\Rightarrow$ **construction**) Let $-$ be an operation over B that satisfies success, inclusion, non-contravention, weak uniformity, pre-expansion and relevance. Let γ be a selection function defined as follows:

$$\gamma(B \perp \neg\alpha) = \{X \in B \perp \neg\alpha : B * \alpha \subseteq X\} \text{ if } B \perp \alpha \neq \emptyset$$
$$= \{B\} \text{ otherwise.}$$

From *weak uniformity* and Lemma 4.6, we have that γ is well defined. It will be proved that (1) if $B \perp \alpha \neq \emptyset$ then $\gamma(B \perp \alpha) \neq \emptyset$ and (2) $B * \alpha = B *_\gamma \alpha$.

1. If $B \perp \neg\alpha \neq \emptyset$ then, by Lemma 4.4, $\neg\alpha \notin Cn(\emptyset)$. By *non-contravention* $\neg\alpha \notin Cn(B*\alpha)$ and, by *upper bound property*, there is X such that $B*\alpha \subseteq X \in B \perp \neg\alpha$. Hence, $\gamma(B \perp \neg\alpha) \neq \emptyset$.

2. If $B \perp \neg\alpha = \emptyset$ then $B *_\gamma \alpha = B + \alpha$. Furthermore, by Lemma 4.4, $\neg\alpha \in Cn(\emptyset)$. By *relevance* if $\beta \in B \setminus B * \alpha$ then there must be a B' such that $\neg\alpha \notin Cn(B')$. Since $\neg\alpha \in Cn(\emptyset)$, this B' cannot exist and there is no $\beta \in B \setminus B * \alpha$. Hence, $B \subseteq B * \alpha$. By *success*, $B + \alpha \subseteq B * \alpha$ and, by *inclusion*, $B * \alpha \subseteq B + \alpha$. We conclude that $B * \alpha = B *_\gamma \alpha = B + \alpha$.

If $B \perp \neg\alpha \notin \emptyset$ then, by *pre-expansion* $B*\alpha \subseteq B *_\gamma \alpha$. Let $\beta \in B *_\gamma \alpha$ and $\beta \notin B*\alpha$. By *pre-expansion*, $\beta \in B + \alpha \setminus (B+\alpha)*\alpha$. Then, by *relevance* there is a set B' such that $(B + \alpha) * \alpha \subseteq B' \subseteq B + \alpha \neg\alpha \notin Cn(B')$ and $\neg\alpha \in Cn(B' \cup \{\beta\})$. It follows, by *upper bound property*, that there is X such that $B' \subseteq X \in B + \alpha \perp \neg\alpha$. Since $B' \subseteq X \neg\alpha \notin Cn(X)$ and $\neg\alpha \in Cn(B' \cup \{\beta\})$, we have that $\beta \notin X \in B + \alpha \perp \neg\alpha$. It follows that $\beta \notin B *_\gamma \alpha$. $\square$

4.3.3 Belief Base Semi-Revision

Revision operations presented in previous section assume that agent must accept the input α. This obligation of accepting the input is called *principle of primacy of the input*. Hansson suggested to drop this principle and considered an operation called *semi-revision* [Han97]. Semi-revision delegates the task of choosing to accept the input to the selection mechanism (selection or incision function). In terms of postulate, the main difference between revision and semi-revision is that the latter does not satisfy success.

In this section we will assume that there is a sentence $\bot$ in the language $\mathscr{L}$ and that a set of sentences A is inconsistent if and only if it derives $\bot$.

Semi-revision is constructed by first adding the input α to the belief base B and then removing inconsistencies (*consolidation*). During consolidation the input may be removed from the belief base depending on the selection mechanism. It will be defined two types of consolidation: kernel and partial meet.

Let σ be a incision function over a belief base B. A kernel semi-revision $?_\sigma$ is defined as follows:

$$B?_\sigma\alpha = (B + \alpha) \setminus \sigma((B + \alpha) \!\perp\!\!\!\perp \bot)$$

Some postulates that characterize semi-revision were already presented. This is the case of inclusion, pre-expansion. Some postulates such as relevance and core-retainment were adapted to deal with consistency instead of non-contravention. Furthermore, there are two new postulates, namely, consistency and internal exchange.

(**relevance**) If $\beta \in B$ and $\beta \notin B?\alpha$ then there is a B' such that $B?\alpha \subseteq B' \subseteq B$ and B' is consistent, but $Cn(B' \cup \{\beta\})$ is not.

(**core-retainment**) If $\beta \in B$ and $\beta \notin B?\alpha$ then there is a B' such that $B' \subseteq B$ and B' is consistent, but $B' \cup \{\beta\}$ is not.

(**internal exchange**) If $\alpha, \beta \in B$ then $B?\alpha = B?\beta$.

(**consistency**) If α is consistent then $B?\alpha$ is consistent.

Representation Theorem 4.29 (KSR) [Han97] *Let $\langle \mathscr{L}, Cn \rangle$ be a logic that satisfies monotonicity, compactness. An operation ? over a belief base B is a kernel semi-revision iff for every α the operation ? satisfies: consistency, inclusion, core-retainment, pre-expansion and internal exchange.*

Proof (**construction $\Rightarrow$ postulates**)

consistency, inclusion, pre-expansion, and internal exchange: Follows from construction.

core-retainment: Let $\beta \in B \setminus B?\alpha$. It follows that $\beta \in \sigma(B + \alpha \!\perp\!\!\!\perp \bot)$. Hence, $\beta \in X$ for some $X \in B + \alpha \!\perp\!\!\!\perp \bot$. Notice that $X \setminus \{\beta\} \subseteq B + \alpha \, X \setminus \{\beta\}$ is consistent and X is not.

(**postulates $\Rightarrow$ construction**) Let ? be an operation over B that satisfies consistency, inclusion, core-retainment, pre-expansion, and internal exchange. Let σ be an incision function defined as follows:

$$\sigma(B \!\perp\!\!\!\perp \bot) = B \setminus \{\beta : \beta \in B?\alpha \text{ for some } \alpha \in B\}$$

It will be proved that (1) σ is an incision function (2) $B?\alpha = B?_\sigma\alpha$.

1. If $\beta \in \sigma(B \!\perp\!\!\!\perp \bot)$ then $\beta \in B \setminus B?\alpha$. It follows, by *core-retainment*, that there is a $B' \subseteq B$ such that B' is consistent and $B' \cup \{\beta\}$ is not. By Lemma 4.21, there is X such that $\beta \in X \in B \!\perp\!\!\!\perp \bot$. Hence, $\beta \in \bigcup B \!\perp\!\!\!\perp \bot$. Let $\emptyset \neq X \in B \!\perp\!\!\!\perp \bot$

and suppose that $X \cap \sigma(B \perp\!\!\!\perp \perp) = \emptyset$. It follows that $X \subseteq \{\beta : \beta \in B?\alpha$ for some $\alpha \in B\}$. Take any $\alpha \in B$ consistent. By *internal exchange*, $X \subseteq B?\alpha$. Since X is inconsistent, $B?\alpha$ is inconsistent, by *monotonicity*, which contradicts the *consistency*. Hence, $X \cap \sigma(B \perp\!\!\!\perp \perp) \neq \emptyset$.

2.

$$\sigma(B + \alpha \perp\!\!\!\perp \perp) = B + \alpha \setminus \{\beta : \beta \in (B + \alpha)?\delta \text{ for some } \delta \in B\}$$
$$= B + \alpha \setminus (B + \alpha)?\alpha \quad \text{by internal exchange}$$
$$= B + \alpha \setminus B?\alpha \quad \text{by pre-expansion}$$

It follows that:

$$B?_\sigma\alpha = B + \alpha \setminus \sigma(B' + \alpha \perp\!\!\!\perp \perp)$$
$$= B + \alpha \setminus (B + \alpha \setminus B?\alpha)$$
$$= B?\alpha \qquad\qquad \text{by inclusion}$$

$\square$

Let γ be a selection function over a belief base B. A partial meet semi-revision $?_\gamma$ is defined as follows:

$$B?_\gamma\alpha = \bigcap((B + \alpha) \perp\!\!\!\perp \perp)$$

Representation Theorem 4.30 (PMSR) [Han97] *Let $\langle \mathscr{L}, Cn \rangle$ be a logic that satisfies monotonicity, compactness. An operation ? over a belief base B is a partial meet semi-revision iff for every α the operation ? satisfies: consistency, inclusion, relevance, pre-expansion and internal exchange.*

Proof (**construction $\Rightarrow$ postulates**)

consistency, inclusion, pre-expansion, and internal exchange: Follows directly from construction.

relevance: If $\beta \in B \setminus B?\alpha$ then there is $X \in \gamma(B + \alpha \perp\!\!\!\perp \perp)$ such that $\beta \notin X$. Of course $B?\alpha \subseteq X \subseteq B + \alpha$, X is consistent and $X \cup \{\beta\}$ is not.

(**postulates $\Rightarrow$ construction**) Let ? be an operation over B that satisfies consistency, inclusion, relevance, pre-expansion, and internal exchange. Let γ be a selection function defined as follows:

$$\gamma(B \perp\!\!\!\perp \perp) = \{X \in B \perp\!\!\!\perp \perp : B?\alpha \subseteq X \text{ for some } \alpha \in B\}$$

It will be proved that (1) γ is a selection function (2) $B?\alpha = B?_\sigma\alpha$.

1. By *consistency*, $B?\alpha$ is consistent. It follows, by *upper bound property*, that there is X such that $B?\alpha \subseteq X \in B \perp\!\!\!\perp \perp$.
2. By *pre-expansion*, $B?\alpha \subseteq B?_\sigma\alpha$. Now let $\beta \notin B?\alpha$. If $\beta \notin B + \alpha$ then $\beta \notin B?_\sigma\alpha$ trivially. So let $\beta \in B + \alpha$. By *pre-expansion* and *relevance*, there is B' consistent such that $B?\alpha \subseteq B' \subseteq B + \alpha$ and $B' \cup \{\beta\}$ is not consistent. It follows, by *upper bound property* that there is X such that $B' \subseteq X \in B + \alpha \perp\!\!\!\perp \perp$. Since X is consistent, $X \subseteq B'$ and $B' \cup \{\beta\}$ is inconsistent, then $\beta \notin X$. Hence, $\beta \notin B?_\gamma\alpha$.

$\square$

Table 4.1 Representation theorems for classical belief revision

Construction	Postulates	Logic properties
Partial meet contraction	AGM postulates for contraction	AGM assumptions
Partial meet revision	AGM postulates for revision	AGM assumptions
Kernel base contraction	Suc, Inc, Uni, CR	Compactness
Partial meet base contraction	Suc, Inc, Uni, Rel	Compactness
External kernel base revision	Suc, Cons, Inc, WU, PE, CR	Compactness and α-local non contr.
External PM base revision	Suc, Cons, Inc, WU, PE, Rel	Compactness and α-local non contr.
Internal kernel base revision	Suc, Cons, Inc, Uni, CR	Compactness and α-local non contr.
Internal PM base revision	Suc, Cons, Inc, Uni, Rel	Compactness and α-local non contr.
Semi-revision	Cons, Inc, PE, IE	Compactness

4.4 Conclusion

Belief revision deals with the dynamics of epistemic states. We presented in this chapter two approaches for belief revision: belief set approach, also called AGM theory, and belief base approach. In the first approach the epistemic state of the agent is represented as a set closed under logical consequence, while in the second the epistemic state is represented as an arbitrary set of sentences.

The dynamics of belief revision is achieved via operations on these sets. Some operations both on belief set and in belief base were considered, namely, expansion, revision, contraction, and semi-revision. Each operation was studied in two aspects: as a set of rationality postulates that they satisfy and as a construction. The main results on the belief revision literature relate these two aspects showing that certain set of rationality postulates fully characterizes certain construction.

Once this result, called representation theorem, is achieved one can focus separately on the postulates to understand the properties of the operation and on the construction to study computational aspects. Representation theorems presented in this chapter assume certain properties on the underlying logic. Representation theorems for belief set operations assume that the logic satisfies AGM assumptions and for belief base revision assumes that the logic is compact and closed under negation. Hence, most of these results do not hold for most logics presented in Chap. 3. Two exceptions are semi-revision and contraction in belief base revision that hold for every compact logic. Table 4.1 sums up the logical property required for each representation theorem of this chapter. The first column of the table indicates the construction, the second column contains the abbreviations of the postulates that characterizes the construction and the last column indicates which logical properties are required by the representation theorem.

The seminal article by Alchourràn Gärdenfors and Makinson [AGM85] is still the main work about AGM theory. More details on classical AGM theory can be found in [Gar88] and in [Han99]. Belief base operations were extensively studied by Hansson in [Han99] and by Wassermann in [Was00].

References

[AGM85]   Carlos Alchourrón, Peter Gärdenfors, and David Makinson On the logic of theory change. *Journal of Symbolic Logic*, 50(2):510–530, 1985.

[AM81]   Carlos Alchourrón and David Makinson. Hierarchies of regulation and their logic. In Hilpinen, editor, *New studies in deontic logic*, pages 125–148. D. Reidel Publishing Company, 1981.

[FPA06]   Giorgos Flouris, Dimitris Plexousakis, and Grigoris Antoniou. On generalizing the AGM postulates. *In Proceedings of the 3rd European Starting AI Researcher Symposium (STAIRS-06)*, 2006.

[Fuh97]   André Fuhrmann. *An Essay on Contraction*. CSLI Publications, Stanford University, USA, 1997.

[Gar88]   Peter Gärdenfors. *Knowledge in Flux: Modeling the Dynamics of Epistemic States*. MIT Press, Cambridge, UK, 1988.

[Han91]   Sven Ove Hansson Belief contraction without recovery. *Studia Logica*, 50(2):251–260, 1991.

[Han93]   Sven Ove Hansson. Reversing the Levi identity. *Journal of Philosophical Logic*, 22:637–669, 1993.

[Han94]   Sven Ove Hansson. Kernel contraction. *Journal Symbolic Logic*, 59(3):845–859, 1994.

[Han97]   Sven Ove Hansson. Semi-revision (invited paper). Journal of Applied Non-Classical Logics, 7(2), 1997.

[Han99]   Sven Ove Hansson. *A Textbook of Belief Dynamics*. Kluwer Academic, 1999.

[Har86]   Gilbert Harman. *Change In View: Principles of Reasoning*. Bradford Books. MIT Press, 1986.

[HW02]   Sven Ove Hansson and Renata Wassermann. Local change. *Studia Logica*, 70(1):49–76, 2002.

[Mak87]   David Makinson. On the status of the postulate of recovery in the logic of theory change. *Journal of Philosophical Logic*, 16:383–394, 1987.

[Neb90]   Bernhard Nebel. *Reasoning and revision in hybrid representation systems*. Springer-Verlag New York, Inc., New York, NY, USA, 1990.

[Pea88]   J. Pearl *Probabilistic Reasoning in Intelligent Systems: Networks of Plausible Inference*. Morgan Kaufmann, 1988.

[RW08]   Márcio M. Ribeiro and Renata Wassermann. Degrees of recovery and inclusion in belief base dynamics. In Maurice Pagnucco and Michael Thielscher, editors, *Proceedings of the 12th International Workshop on Non-Monotonic Reasoning (NMR 2008)*, Sidney, Australaia, Setembro 13–15, 2008, Sidney, Australia, Setembro 13–15 2008.

[Sta84]   Robert C. Stalnaker. *Inquiry*. The MIT Press, Cambridge, MA, 1984.

[Was00]   Renata Wassermann. *Resource Bounded Belief Revision*. PhD thesis, Universiteit van Amsterdan, Janeiro 2000.

[Woj88]   R. Wójcicki. *Theory of logical calculi: basic theory of consequence operations*. Synthese library. Kluwer Academic Publishers, 1988.

Chapter 5
AGM Contraction in Non-Classical Logics

Abstract Recovery is certainly the most controversial AGM postulate. Since the first publications in the area many works criticize several aspects of recovery. In order to avoid the undesirable consequences of recovery, Hansson proposes to exchange it by a postulate called *relevance*. However, in classical logics relevance and recovery are equivalent. In this chapter, we defend the use of relevance instead of recovery in non-classical logics for mainly three reasons: relevance deals more directly with minimality, it is compliant with a wider class of logics, and it better characterizes partial meet contraction.

Keywords Contraction · Relevance · Recovery · Partial meet contraction · Minimality · AGM-compliance · Non-classical logics

This chapter presents contraction in belief sets. Our main goal is to show how to apply this theory in non-classical logics.

Recovery is the most controversial AGM postulate [Mak87], [Han91], [FPA06], [Lev91], [Nie91]. The oldest and most commented criticism is from Makinson. Makinson argues that recovery is the only postulate which is unnecessary to construct a revision via Levi identity [Mak87]. He argues that recovery should be dropped and defines an operation called *withdrawal*. A withdrawal satisfies all the AGM postulates for contraction but recovery.

Hansson rejects withdrawal arguing that dropping recovery altogether is too drastic [Han91]. He agrees, though, that the postulate is unintuitive presenting the following example:

Example 5.1 "I believe that 'Cleopatra had a son' α_1 and that 'Cleopatra had a daughter' α_2, hence, I also believe that 'Cleopatra had a child' $\alpha_1 \vee \alpha_2$. Then I receive the information that makes me give up my belief in $\alpha_1 \vee \alpha_2$, and make me contract my belief accordingly $K - \alpha_1 \vee \alpha_2$. After that I learn from a reliable source that Cleopatra did have a child. It seems plausible to me to add $\alpha_1 \vee \alpha_2$ to my belief base without neither adding α_1 nor α_2 to it."

M. M. Ribeiro, *Belief Revision in Non-Classical Logics*,
SpringerBriefs in Computer Science,
DOI: 10.1007/978-1-4471-4186-0_5, © The Author(s) 2013

What Hansson proposes, instead of simply dropping recovery, is to exchange it with some other, more intuitive, minimality postulate. For this purpose *relevance* was introduced:

(**relevance**) If $\beta \in K$ and $\beta \notin K - \alpha$ then there is a K' such that $K - \alpha \subseteq K' \subseteq K$, $\alpha \notin Cn(K')$, but $\alpha \in Cn(K' \cup \{\beta\})$.

Relevance guarantees minimality of contraction preventing the removal of sentences which are irrelevant to infer the input α. Hansson, however, proved that relevance and recovery are equivalent in the presence of AGM assumptions and the other contraction postulates. He concludes the paper claiming that recovery should be accepted as an "emerging property, rather than as a fundamental postulate, of belief set contraction".

Another criticism against the use of recovery as a postulate for contraction can be found in [FPA06]. In this article, it is showed that AGM postulates are not applicable to a wide class of non-classical logics and that recovery is the main reason behind this failure.

The main goal of this chapter is to show that relevance is indeed a good replacement for recovery. To facilitate the exposure we call *relevance-contraction* the operation that satisfies the AGM postulates for contraction with recovery exchanged by relevance. We defend that this is a good choice of postulates for non-classical logics with three arguments:

1. *compliance:* relevance-contraction postulates are compliant with a wide class of logics.
2. *equivalence with partial meet:* relevance-contraction postulates are equivalent to partial meet in a wide class of logics.
3. *rationality:* although relevance and recovery are equivalent in many logics, this is not the case in general.

Next section presents a straightforward generalization of the AGM postulates for the case where the input is a set of sentences. Section 5.2 shows that most logics are not compliant with AGM postulates. Section 5.3 shows that every compact logic is compliant with AGM postulates with recovery exchanged by relevance. Section 5.4 proves a representation theorem that characterizes partial meet contraction in any compact logic. Section 5.5 shows that in certain logics relevance and recovery are equivalent while in others they are not. Finally, Sect. 5.6 concludes the chapter.

5.1 Generalized AGM Postulates

In *CPL* every finite set $a_1, \ldots, a_n$ is equivalent to a single sentence $a_1 \wedge \cdots \wedge a_n$, i.e., $C_{CPL}(\{a_1, \ldots, a_n\}) = C_{CPL}(a_1 \wedge \cdots \wedge a_n)$. However, in some logics that we are interested in, such as Description Logics, it is not possible to represent finite sets with a single sentence. For this reason it is convenient to generalize contraction in order for it to accept finitely representable sets of sentences as input. Such generalized operation is called multiple contraction in [FH94].

This generalization was first proposed by Fuhrmann and Hansson in [FH94] and then defined by Flouris in [Flo06]. Notice that it is a straightforward generalization of the AGM postulates.

Let K be a belief set and A be a finitely representable set over $\mathscr{L}$. The generalized AGM postulates for contraction are the following:

(closure) $K - A = Cn(K - A)$.
(success) If $A \nsubseteq Cn(\emptyset)$ then $A \nsubseteq K - A$.
(inclusion) $K - A \subseteq K$.
(vacuity) If $A \nsubseteq K$ then $K - A = K$.
(recovery) $K \subseteq (K - A) + A$.
(extensionality) If $Cn(A) = Cn(B)$ then $K - A = K - B$.

Although straightforward, this generalization is not the only one found in the literature. The set of postulates presented above interpret the input conjunctively, i.e., the contraction $K - A$ should remove at least one element of A (see the postulate of success). Alternatively, we could interpret the input set disjunctively. In this case $K - A$ should remove every element of A. In [FH94] the former operation is called *choice contraction* while the later is called *package contraction*. In our approach only choice contraction will be considered.

In the previous section relevance was presented. It is convenient to present a generalization of this postulate also.

(relevance) If $\beta \in K$ and $\beta \notin K - A$ then there is a K' such that $K - A \subseteq K' \subseteq K$, $A \nsubseteq Cn(K')$, but $A \subseteq Cn(K' \cup \{\beta\})$.

5.2 AGM Compliance

In Sect. 4.2 postulates for AGM contraction were presented and it was shown that they fully characterize a construction called partial meet contraction. The equivalence between AGM postulates and partial meet contraction holds for logics that satisfies certain conditions called AGM assumptions. However, most logics listed in Chap. 3 do not satisfy these assumptions.

In [FPA05] the authors argue that AGM postulates for contraction are not even applicable to certain logics. Consider the logic from Example 2.3:

$$\mathscr{L} = \{a, b\}$$
$$Cn(\mathscr{L}) = Cn(\{b\}) = \mathscr{L}$$
$$Cn(\{a\}) = \{a\}$$
$$Cn(\emptyset) = \emptyset$$

Let $K = \mathcal{L}$ and $A = \{a\}$. By inclusion, success and closure the only candidate for $K - A$ in this logic is $\emptyset$. However, if $K - A = \emptyset$ we would have $(K - A) + A = \{a\} \neq K$, i.e., recovery would not be satisfied.

This is an example of a logic which is not AGM-compliant. As we will see most logics listed in Chap. 3 are not AGM-compliant.

Definition 5.2 (*AGM-compliance*) [FPA05] A logic $\langle \mathcal{L}, Cn \rangle$ is *AGM compliant* iff for every belief set K there is at least one operation $-$ over K that satisfies the generalized AGM postulates for contraction.

As illustrated by Example 2.3, not every Tarskian logic is AGM compliant. In [FPA05] the authors showed that the necessary and sufficient condition for AGM compliance of a logic is decomposability:

Theorem 5.3 [FPA05] *A logic* $\langle \mathcal{L}, Cn \rangle$ *is AGM compliant iff it is decomposable.*

Proof ($\Rightarrow$) Let K be a belief set and A be a finitely representable set of sentences in a logic $\langle \mathcal{L}, Cn \rangle$ and let $Cn(\emptyset) \subset Cn(A) \subset K$. We will show that there is a set K' such that (1) $Cn(K') \subset K$ and (2) $Cn(K' \cup A) = K$.

1. Let $-$ be an AGM contraction over K. Since $\langle \mathcal{L}, Cn \rangle$ is AGM compliant, $-$ is defined for every finitely representable set A. Now let $K' = K - A$.
   By *inclusion* $K' \subseteq K$ and, since $Cn(\emptyset) \subset Cn(A)$, by *success* $A \not\subseteq K$. Hence $K' \neq K$.
2. $K \subseteq K' + A$ follows directly from *recovery*. Furthermore, since $Cn(A) \subset Cn(K)$ and, by *inclusion*, $K' \subseteq K$, we have that $K' + A \subseteq K$.

($\Leftarrow$) Let K be a belief set and A be a finitely representable set in a logic $\langle \mathcal{L}, Cn \rangle$. Consider a function γ that given a non-empty set $K^-(A)$ returns one element of $K^-(A)$, i.e., $\gamma(K^-(A)) \in K^-(A)$. Let us define an operation $-$ over K as follows:

$$
\begin{aligned}
K - A &= Cn(\gamma(K^-(A)) && \text{if } Cn(\emptyset) \subset Cn(A) \subset K \\
&= K && \text{if } Cn(\emptyset) = Cn(A) \text{ or } A \not\subseteq K \\
&= Cn(\emptyset) && \text{if } Cn(A) = K
\end{aligned}
$$

Notice that $K^-(A) \neq \emptyset$, because the underlying logic is decomposable. The operation $-$ defined above is an AGM contraction.

closure, inclusion, vacuity: Follows directly from construction.

extensionality: If $Cn(A) = Cn(B)$ then it is not difficult to notice that $K^-(A) = K^-(B)$. Since γ is a function, $\gamma(K^-A) = \gamma(K^-B)$. Hence, $K - A = K - B$.

success: If $A \not\subseteq Cn(\emptyset)$ then we have two cases.
If $A = K$ then $K - A = Cn(\emptyset)$ and, since $A \not\subseteq Cn(\emptyset)$, the success is satisfied.
If $A \neq K$ then $K - A = K'$ such that $Cn(K') \subset K$ and $K' + A = K$.
Now suppose that $A \subseteq Cn(K')$ then $K' + A = K'$ which is contradiction.
Hence, $A \not\subseteq Cn(K')$.

recovery: If $Cn(A) = Cn(\emptyset)$ or $A \not\subseteq K$ then $K - A = K$, hence, $K \subseteq (K - A) + A$.
If $Cn(A) = K$ then $K - A = Cn(\emptyset)$ and $(K - A) + A = Cn(A)$, hence, $K \subseteq (K - A) + A$.

Finally, if $Cn(\emptyset) \subset Cn(A) \subset K$ then $K - A = K'$ for some $K' \in K^-(A)$ and, hence, $K' + A = K$. We conclude that $K \subseteq (K - A) + A$. □

From this result we have as corollary the following:

Corollary 5.4 *All Boolean logics are AGM compliant.*

Proof It follows directly from Theorem 5.3 and Proposition 2.10. □

5.3 Relevance Compliance

Chapter 3 listed several logics such as intuitionistic logic, Horn logic, and many DLs which are not decomposable and, hence, not AGM compliant. Are these logics compliant with relevance, i.e., are they compliant with the postulates for relevance contraction? In other words, are this logics relevance-compliant?

Definition 5.5 (*relevance-compliance*) A logic $\langle \mathcal{L}, Cn \rangle$ is *relevance-compliant* iff for every belief set K there is at least one operation $-$ over K that satisfies the generalized postulates for relevance-contraction.

The following theorem shows that compactness is a sufficient condition for a logic to be relevance-compliant.

Theorem 5.6 *Every compact logic is relevance-compliant.*[1]

Proof Let K be a belief set and A be a finitely representable set A in a compact logic $\langle \mathcal{L}, Cn \rangle$. Consider a function γ that given a non empty set $K \perp A$ returns one element of it, i.e., $\gamma(K \perp A) \in K \perp A$. Let $-$ be an operation over K defined as follows:

$$K - A = K \qquad \text{if } A \subseteq Cn(\emptyset)$$
$$= \gamma(K \perp A) \quad \text{otherwise}$$

Notice that when $A \not\subseteq Cn(\emptyset)$, by Lemma 5.9, $K \perp A \neq \emptyset$. Let us show that $-$ is a relevance-contraction:

inclusion: Follows directly from construction.
extensionality: If $Cn(A) = Cn(B)$ then $K \perp A = K \perp B$, by Lemma 5.13. Since γ is a function then $\gamma(K \perp A) = \gamma(K \perp B)$.
closure: Follows from Lemma 5.10.
vacuity: If $A \not\subseteq K$ then $K - A = K' \in K \perp A = \{K\}$. Hence, $K - A = K$.
success: If $A \not\subseteq Cn(\emptyset)$ then $K - A = K' \in K \perp A$. Hence, $A \not\subseteq K - A$.
relevance: If $A \subseteq Cn(\emptyset)$ then $K - A = K$ and relevance is vacuously satisfied.

If $A \not\subseteq Cn(\emptyset)$ then $K - A = K' \in K \perp A$. Let $\beta \in K$ such that $\beta \notin K - A$. Of course $K - A \subseteq K' \subseteq K$ and $A \not\subseteq Cn(K')$. Furthermore, since K' is maximal, $A \subseteq Cn(K' \cup \{\beta\})$ □

---

[1] This theorem uses some lemmas presented in the following section.

Since all logics from Chap. 3 are compact, they are all relevance-compliant. However, as shown in the following example, not every logic is relevance-compliant.

Example 5.7 The logic below is AGM-compliant but not relevance-compliant. It is also closed under negation. The logic $\langle \mathscr{L}, Cn \rangle$ is defined as follows:

$$\mathscr{L} = \{a, x_i, y_j \mid i, j = 1, 2, \ldots\}$$
$$Cn(a) = \{a\}$$
$$Cn(y_1) = \{x_1, y_1\}$$
$$Cn(y_i) = \{y_i\} \text{ for } i > 1$$
$$Cn(x_i) = \{x_j, y_j \mid j \leq i\}$$
$$Cn(X) = \mathscr{L} \text{ if } |X| > 1 \text{ and } a \in X$$
$$Cn(X) = Cn(x_i) \text{ if } X \text{ is finite, } |X| > 1, \quad a \notin X$$
$$\text{and } x_i \in X \text{ or } y_i \in X$$
$$\text{and there is no } j > i \text{ such that } x_j \in X \text{ or } y_j \in X$$
$$Cn(X) = \mathscr{L} \text{ if } X \text{ is infinite}$$
$$Cn(\emptyset) = \emptyset$$

To verify that this logic is AGM-compliant just check each pair *K, A* and identify one contraction result that satisfies the AGM postulates.

The logic is not relevance-compliant. There is no candidate set that satisfies relevance and success for the operation $\mathscr{L} - \{a\}$. If X is infinite then it would not satisfy success. Let X be finite then there is a x_i such that $x_i \notin Cn(X)$. By relevance there must be a K' such that $a \notin Cn(K')$, but $a \in Cn(K' \cup \{x_i\})$. By monotony and the fact that $a \notin K'$ we have that K' must be finite. In this case, $a \notin Cn(K' \cup \{x_i\})$.

5.4 Relevance and Partial Meet Contraction

We already showed that if the underlying logic satisfies the AGM assumptions then partial meet contraction is equivalent to the AGM postulates. This result is central to the AGM theory and it is called representation theorem for partial meet contraction.

We also showed that AGM postulates are not compliant with several interesting logics and relevance was proposed as a replacement for recovery.

The main result in this section shows that the postulates for relevance-contraction fully characterizes partial meet contraction for a wider class of logics. More precisely partial meet contraction is fully characterized by relevance-contraction postulates in every compact logic.

Before showing this result let us generalize remainder set to accept finitely representable sets as input:

Definition 5.8 (*generalized remainder set*) Let K be a belief set and A and finitely representable set of sentences. The remainder set $K \perp A$ is the set such that $X \in K \perp A$ iff:

1. $X \subseteq K$ (X is a subset of K).
2. $A \not\subseteq Cn(K)$ (that does not imply A).
3. if $X \subset X' \subseteq K$ then $A \subseteq Cn(X')$ (and is maximal).

The proof of the representation theorem we are about to enunciate depends on the following lemmas:

Lemma 5.9 (generalized upper-bound property) *If $\langle \mathscr{L}, Cn \rangle$ is compact then for every $K \subseteq \mathscr{L}$, every $X \subseteq K$ and every finitely representable $A \subseteq \mathscr{L}$ for which $A \not\subseteq Cn(X)$, there is a X' such that $X \subseteq X'$ and $X' \in K \perp A$.*

Proof Arrange the elements of K in a sequence $\beta_1, \beta_2, \ldots$. Let $X_0 = X$ and for all $i \geq 1$ defined X_i as follows:

$$X_i = \begin{cases} X_{i-1} & \text{if } A \subseteq Cn(X_{i-1} \cup \{\beta_i\}) \\ X_{i-1} \cup \{\beta_i\} & \text{otherwise.} \end{cases}$$

Let $K' = \bigcup_i X_i$. Since A is finitely representable, there is A' finite which is equivalent to A. If $A \subseteq Cn(K')$, by Lemma 2.2, then $A' \subseteq Cn(K')$. Suppose without loss of generality that $A' = \{a_1, \ldots, a_n\}$. By compactness for every a_j with $1 \leq j \leq n$ there is a finite K_j such that $K_j \subseteq K'$ and $a_j \in Cn(K_j)$. It follows that $A' \subseteq Cn(\bigcup_j K_j)$ and, since $\bigcup_j K_j$ is finite, $A \subseteq Cn(X_i)$ for some i. By Lemma 2.2 $A \subseteq Cn(X_i)$ which is a contraction. It follows that $A \not\subseteq Cn(K')$.

Furthermore, if $K' \subset K'' \subseteq K$ then there is $\beta \in K$ such that $\beta \in K''$ and $\beta \notin K'$. By construction we have that $A \subseteq Cn(K' \cup \{\beta\}$, hence, $A \subseteq Cn(K'')$.

Of course $X \subseteq K' \subseteq K$ which concludes the proof that $X \subseteq K' \in K \perp A$. $\square$

Lemma 5.10 *Let K be a belief set and A a finitely representable set of sentences in a logic $\langle \mathscr{L}, Cn \rangle$. If $X \in K \perp A$ then X is closed under logical consequence, i.e., $X = Cn(X)$.*

Proof If $\beta \in Cn(X)$ then $A \not\subseteq Cn(X \cup \{\beta\})$ and, since X is maximal, $\beta \in X$. $\square$

Lemma 5.11 *Let $\langle \mathscr{L}, Cn \rangle$ be a compact logic. Let K be a belief set and A be finitely representable set of sentences in a logic $\langle \mathscr{L}, Cn \rangle$. Then $K \perp A = \emptyset$ iff $A \subseteq Cn(\emptyset)$.*

Proof ($\Rightarrow$) If $A \not\subseteq Cn(\emptyset)$ then by *generalized upper bound property* (Lemma 5.9) there is $X \in K \perp A$.

($\Leftarrow$) If $A \subseteq Cn(\emptyset)$ then there is no $X \subseteq K$ such that $A \not\subseteq Cn(X)$. $\square$

Lemma 5.12 *Suppose that $\langle \mathscr{L}, Cn \rangle$ is compact and let K be a belief set. Then $K \perp A = K \perp B$ iff for every $X \subseteq K$ it holds that $A \subseteq Cn(X)$ iff $B \subseteq Cn(X)$.*

Proof Consider a set $X \subseteq K$ and suppose that $A \subseteq Cn(X)$ and $B \not\subseteq Cn(X)$. By *generalized upper-bound property* (Lemma 5.9) there is X' such that $X \subseteq X'$

$\in K \perp B$. Since $A \subseteq Cn(X)$ then, by monotonicity, $A \subseteq Cn(X')$. However, in this case, $X' \notin K \perp A$ which contradicts the hypothesis. The converse is trivial.   □

Lemma 5.13 *Let K be a belief set and let A and B be finitely representable sets of sentences in a compact logic* $\langle \mathcal{L}, Cn \rangle$ *then the following hold*:

1. *If* $Cn(A) = Cn(B)$ *then* $K \perp A = K \perp B$.
2. *If* $A \subseteq K$ *and* $K \perp A = K \perp B$ *then* $Cn(A) = Cn(B)$.

Proof

1. If $K' \subseteq K$ then $A \subseteq Cn(K')$ iff $B \subseteq Cn(K')$, because $Cn(A) = Cn(B)$. It follows, by the previous lemma that $K \perp A = K \perp B$.
2. If $A \subseteq K$ then $Cn(A) \subseteq K$ by *monotonicity*. By the previous lemma, $B \in Cn(A)$. Analogously, if $B \subseteq K$ then $A \in Cn(B)$ and, hence, $Cn(A) = Cn(B)$.
   If $B \nsubseteq K$ then $K \perp B = \{K\}$ and, since $K \perp A = K \perp B$, we have that $A \nsubseteq K$.
   □

With these lemmas we are able to prove the following representation theorem:

Representation Theorem 5.14 *Let* $\langle \mathcal{L}, Cn \rangle$ *be a compact logic, let K be a belief set and A be a finitely representable set.* $K - A$ *satisfies the postulates for relevance-contraction iff* $K - A = \bigcap \gamma(K \perp A)$ *for some selection function* γ.

Proof (**construction** $\Rightarrow$ **postulates**)

inclusion: Follows directly from construction.
closure: Follows from Lemma 5.10 and the fact that the intersection of closed sets is closed.
extensionality: Follows from Lemma 5.13.
vacuity: If $A \nsubseteq K$ then $K \perp A = \{K\}$. It follows that $\gamma(K \perp A) = K$.
success: If $A \nsubseteq Cn(\emptyset)$ then, by Lemma 5.11, $K \perp A \neq \emptyset$. Hence $A \nsubseteq \bigcap \gamma(K \perp A)$.
relevance: If $\beta \in K \setminus K - A$ then there is $X \in \gamma(K \perp A)$ such that $\beta \notin X$. By definition, $K -_\gamma A \subseteq X \subseteq K$, $A \nsubseteq Cn(X)$ and $A \subseteq Cn(X \cup \{\beta\})$.

(**postulates** $\Rightarrow$ **construction**)

Let $-$ be an operator satisfying the six postulates and define the selection function γ as follows:

$$\gamma(K \perp A) = \{X \in K \perp A : K - A \subseteq X\} \text{ if } K \perp A \neq \emptyset$$
$$= \{K\} \text{ otherwise.}$$

We have to prove that (1) γ is well defined, (2) γ is a selection function, and (3) $K -_\gamma A = K - A$.

1. If $K \perp A = K \perp B$ then, by Lemma 5.13, $K - A = K - B$. It follows that $\gamma(K \perp A) = \gamma(K \perp B)$
2. If $K \perp A \neq \emptyset$ then, by Lemma 5.11, $A \nsubseteq Cn(\emptyset)$. By *success*, $A \nsubseteq K - A$ and, by Lemma 5.9, there is X such that $K - A \subseteq X \in K \perp A$. Hence, $\gamma(K \perp A) \neq \emptyset$.

3. If $A \subseteq Cn(\emptyset)$ then $K -_\gamma A = K$. Furthermore, by *relevance*, if $\beta \in K \setminus K - A$ then there must be a K' such that $A \not\subseteq Cn(K')$. Since $A \subseteq Cn(\emptyset)$, this K' cannot exist. It follows that there is no $\beta \in K \setminus K - A$ i.e. $K \subseteq K - A$. By *inclusion*, $K - A \subseteq K$, hence, $K - A = K -_\gamma A = K$.

If $A \not\subseteq Cn(\emptyset)$ then, by construction $K - A \subseteq K -_\gamma A$. Now let $\beta \notin K - A$. If $\beta \notin K$ then $\beta \notin K -_\gamma A$ trivially. So let $\beta \in K$. By *relevance*, there is K' such that $K - A \subseteq K' \subseteq K$, $A \not\subseteq Cn(K')$ and $A \subseteq Cn(K' \cup \{\beta\})$. It follows, by Lemma 5.9, that there is X such that $K' \subseteq X \in K \perp A$. Since $K' \subseteq X$, $A \subseteq Cn(K' \cup \{\beta\})$ and $A \not\subseteq Cn(X)$, we have that $\beta \notin X$. Hence, $\beta \notin \bigcap \gamma(K \perp A)$ □

5.5 Rationality

It was showed that relevance is compliant with every compact logic and, furthermore, in these logics relevance-contraction postulates fully characterize partial meet contraction. This chapter will explore yet another aspect of this set of postulates: its rationality.

First, notice that relevance tries to capture the minimality intuition directly. It states that no sentences can be removed from the belief set unless it "helps" to derive the input. Although much less elegant, relevance treats minimality much more directly than recovery.

Second, relevance is not that different from recovery. In fact, as noticed by Hansson, recovery and relevance are equivalent in every logic that satisfies the AGM assumptions. Since partial meet contraction is fully characterized by AGM postulates, we have that if a logic satisfies the AGM assumptions then relevance-contraction postulates, AGM postulates, and partial meet contraction are all equivalent.

It is not difficult to show that the equivalence between relevance and recovery can be generalized to Boolean logics.

Lemma 5.15 *Let K be a belief set and A be a finitely representable set of sentences in a Boolean logic $\langle \mathcal{L}, Cn \rangle$. If an operation $-$ over K satisfies recovery and $\beta \in K$ then for every complement A' of A, $Cn(A') \cup Cn(\beta) \subseteq K - A$.*

Proof If $\beta \in K$ then, by *recovery*, $\beta \in (K - A) + A$. Let A' be a complement of A. Since $A' \subseteq (K - A) + A'$, we have that $Cn(\beta) \cap Cn(A') \subseteq ((K - A) + A) \cap ((K - A) + A')$. It follows, by *distributivity*, that $Cn(\beta) \cap Cn(A') \subseteq K - A + (Cn(A) \cap Cn(A'))$. Hence, $Cn(\beta) \cap Cn(A') \subseteq K - A$. □

Theorem 5.16 *If a logic $\langle \mathcal{L}, Cn \rangle$ is Boolean, K is a belief set and A is finitely representable then relevance and recovery are equivalent (in the presence of the other AGM postulates).*

Proof (**relevance $\Rightarrow$ recovery**) Let $\beta \notin (K - A) + A$ and suppose that $\beta \in K$. Take A' a complement of A (which exists because the logic is complemented) and let $X = Cn(A') \cap Cn(\beta)$. It follows that $X + A = Cn(Cn(A') \cap Cn(\beta) \cup A)$ and, by *distributivity*, this is equal to $Cn(A \cup A') \cap Cn(A \cup \{\beta\}) = Cn(A \cup \{\beta\}$. Hence,

$\beta \notin X + A$. Since $\beta \notin (K - A) + A$, we have that $X \nsubseteq K - A$ i.e., there is $\beta \in X \subseteq K$ such that $\beta \notin K - A$. It follows, by *relevance*, that there is K' such that $K - A \subseteq K' \subseteq K$, $A \nsubseteq Cn(K')$ and $A \subseteq Cn(K' \cup \{\beta\})$. Since $\beta \in X$, we have that:

$$A \subseteq Cn(K' \cup X)$$
$$= Cn(K' \cup (Cn(A') \cap Cn(\beta)))$$
$$= Cn(K' \cup A') \cap Cn(K' \cup \{\beta\}) \quad \text{by distributivity}$$
$$\subseteq Cn(K' \cup A')$$

Furthermore, $A \subseteq Cn(K' \cup A)$, hence, $A \subseteq Cn(K' \cup A) \cap Cn(K' \cup A')$. By *distributivity*, $A \subseteq Cn(K' \cup (Cn(A) \cap Cn(A'))) = Cn(K')$. However, this contradicts the definition of K'. We conclude that $\beta \notin K$.

(recovery $\Rightarrow$ relevance)

If $A \nsubseteq K$ then, by *vacuity* $K - A = K$ and relevance is vacuously satisfied. If $A \subseteq K$ then let $\beta \in K \setminus K - A$ and B' be a complement of β. Now let $K' = Cn(K - A \cup (Cn(A) \cap Cn(B')))$. We will show that (1) $K - A \subseteq K' \subseteq K$, (2) $A \subseteq Cn(K' \cup \{\beta\}$ and (3) $A \nsubseteq Cn(K')$.

1. By construction, $K - A \subseteq K'$ and, since $A \subseteq K$, $K' \subseteq K$, by *inclusion*.
2. By *distributivity*, $Cn(K' \cup \{\beta\}) = Cn(K - A \cup B' \cup \{\beta\}) \cap Cn(K - A \cup A \cup \{\beta\}$. It follows that $Cn(K' \cup \{\beta\} = Cn(K - A \cup A \cup \{\beta\})$ and, hence, $A \subseteq Cn(K' \cup \{\beta\})$.
3. Suppose that $A \subseteq Cn(K') = Cn(K - A \cup (Cn(A) \cap Cn(B')))$. Since $\beta \in Cn(K - A \cup \{\beta\})$, we have that $Cn(A) \cap Cn(\beta) \subseteq Cn(K - A \cup B') \cap Cn(K - A \cup \{\beta\})$. It follows, by *distributivity*, that $Cn(A) \cap Cn(\beta) \subseteq Cn(K - A \cup (Cn(B' \cap Cn(\beta))))$ and, hence, $Cn(A) \cap Cn(\beta) \subseteq Cn(K - A)$.
   Furthermore, by Lemma 5.15 and the fact that the logic is complemented, there is a complement A' of A and $Cn(A') \cap Cn(\beta) \subseteq K - A$. From this and $Cn(A) \cap Cn(\beta) \subseteq Cn(K - A)$, we have that $Cn((Cn(A) \cap Cn(\beta)) \cup (Cn(A) \cap Cn(\beta)) \subseteq Cn(K - A)$ and, by *distributivity*, $Cn(\{\beta\} \cup (Cn(A) \cap Cn(A')) \subseteq Cn(K - A)$. Hence, by *closure*, $\beta \in K - A$ which contradicts the definition of β. We conclude that $A \nsubseteq K'$.   □

Although for Boolean logics relevance and recovery are equivalent in the presence of the other AGM postulates, this is not the case in general. In several interesting logics relevance and recovery are completely distinct postulates. Example 5.17 shows a logic which is both AGM-compliant and relevance-compliant such that neither relevance implies recovery nor the converse.

Example 5.17 This example shows a simple logic which is both AGM-compliant and relevance-compliant, but for which relevance does not imply recovery or vice versa; also, in this logic partial meet contraction does not give the same results as the AGM postulates with recovery. The logic $\langle \mathscr{L}, Cn \rangle$ is defined as follows:

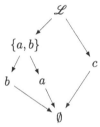

Fig. 5.1 Logic of example 5.17

$$\mathscr{L} = \{a, b, c\}$$
$$Cn(\{b, c\}) = Cn(\{a, c\}) = \mathscr{L}$$
$$Cn(\{a, b\}) = \{a, b\}$$
$$Cn(a) = \{a\}$$
$$Cn(b) = \{b\}$$
$$Cn(c) = \{c\}$$
$$Cn(\emptyset) = \emptyset$$

The logic is finite, thus compact, so it is also relevance-compliant. It is trivial to see that it is also AGM-compliant.

Consider now the operation $\mathscr{L} - \{a, b\}$. Note that if $\{a\}$ is accepted as the result of this operation, relevance is satisfied, but recovery is not. If $\{c\}$ is accepted as the result, then recovery is satisfied, but relevance is not. Hence, neither relevance implies recovery, nor the converse.

Figure 5.1 shows a diagram with the lattice that illustrates this logic.

This example shows that in non-classical logics, even in logics that are relevance-compliant and AGM-compliant, relevance and recovery are independent. Hence, there is no need to accept recovery as an "emerging property" of contraction.

5.6 Conclusion

Most results in the previous chapter hold only for logics that satisfy certain properties. In particular, representation theorem for contraction in belief sets hold in logics that satisfy AGM assumptions which is not the case of many logics listed in Chap. 3.

In this chapter we showed that the problem is even bigger. Certain logics are not even compliant with AGM postulates for contraction. Those logics coincide with non-decomposable logics which include Horn logic, Intuitionistic logic, and many Description Logics, for example.

To avoid problems of compliance, it was suggested the exchange of recovery by a postulate called relevance. In the rest of the chapter several results were presented to defend the set of rationality that contains AGM postulates with recovery exchanged by relevance.

This set of postulates deals with minimality more directly. It is compliant with every compact logic which includes, in particular, every logic presented in Chap. 3 (see Theorem 5.6). It better characterizes partial meet contraction, i.e., it characterizes this construction in a wider range of logics (Representation Theorem 5.14). Furthermore, although it is equivalent with AGM postulates in Boolean logics (Theorem 5.16), there are several logics such that relevance and recovery are distinct (see Example 5.17). In these logics recovery and relevance represent are completely distinct concepts and relevance do not need to be accepted as an emerging property of contraction.

Relevance captures the notion of minimality while recovery captures the notion of recoverability. It only happens that in well-behaved logics these notions coincide. In most other logics this is not the case.

Results about AGM compliance are from Flouris at al. [FPA05], [FPA06], [Flo06]. The other results are from works co-authored by Wassermann [RW06], [RW10] and works co-authored by Wassermann, Flouris and Antoniou to appear.

References

[FH94]   André Fuhrmann and Sven Ove Hansson. A survey of multiple contractions. *Journal of Logic, Language and Information*, 3(1):39–75, Janeiro 1994.

[Flo06]   Giorgos Flouris. *On Belief Change and Ontology Evolution*. PhD thesis, University of Crete, 2006.

[FPA05]   Giorgos Flouris, Dimitris Plexousakis, and Grigoris Antoniou. On applying the AGM theory to DLs and OWL. In Enrico Motta Yolanda Gil, V. Richard Benjamins, and Mark A. Musen, editors, *Proceedings of the 4th International Semantic Web Conference (ISWC 2005)*, pages 216–231, Galway, Ireland, November, 6-10 2005. Springer.

[FPA06]   Giorgos Flouris, Dimitris Plexousakis, and Grigoris Antoniou. On generalizing the AGM postulates. In *Proceedings of the 3rd European Starting AI Researcher Symposium (STAIRS-06)*, 2006.

[Han91]   Sven Ove Hansson. Belief contraction without recovery. *Studia Logica*, 50(2):251–260, 1991.

[Lev91]   Isaac Levi. *The Fixation of Belief and its Undoing: Changing Beliefs through Inquiry.* Cambridge University Press, Outubro 1991.

[Mak87]   David Makinson. On the status of the postulate of recovery in the logic of theory change. *Journal of Philosophical Logic*, 16:383–394, 1987.

[Nie91]   Reinhard Niederée. Multiple contraction. a further case against gärdenfors' principle of recovery. In *Proceedings of the Workshop on The Logic of Theory Change*, pages 322–334, London, UK, 1991. Springer-Verlag.

[RW06]   Márcio Moretto Ribeiro and Renata Wassermann. First steps towards revising ontologies. In *Proceedings of the 2nd Workshop on Ontologies and their Applications (WONTO 2006)*, 2006.

[RW10]   Márcio Moretto Ribeiro and Renata Wassermann. More about AGM revision in description logics. In *Proceedings of the 2nd Workshop Automated Reasoning about Context and Ontology Evolution (ARCOE 10)*, Lisboa, Portugal, Agosto, 16-17 2010.

Chapter 6
AGM Revision in Logics Without Negation

Abstract AGM revision was defined via Levi identity. Levi identity, however, assumes that the underlying logic is closed under negation. Many non-classical logics with great philosophical and computational interest, such as Horn logics and Description Logics, however, are not closed under negation. If we want to use belief revision techniques in these logics we need a way to define revision that doesn't depend on negation. In this chapter we define a construction for revision in belief sets that doesn't depend on negation. This construction is characterized by AGM postulates in well behaved logics. Furthermore, it is presented a set of postulates that characterizes this construction in any distributive and compact logic.

Keywords Belief set revision · Negation · Non-classical logics · Distributivity · Representation theorems

In Sect. 4.2.2 it was showed how to define revision over belief sets via Levi identity using an AGM contraction ($K * \alpha = (K - \neg\alpha) + \alpha$). Levi identity, however, uses the negation of a sentence α in its formulation. The generalization of Levi identity to sets of sentences is not trivial. The main candidate for the negation of a set of sentences A is the complement of A. However, several logics listed in Chap. 3 are not complemented.

In this Chapter it is presented how to construct revision over belief sets in logics that are not closed under negation. As we did previously, the operations are defined over a belief set K and the input is a finitely representable set of sentences A. Besides presenting the construction, postulates that characterize the construction are presented. If the underlying logic satisfies the AGM assumptions then the construction is characterized by AGM postulates for revision. However, if some of these assumptions are dropped then other postulates are needed.

Next section presents a straightforward generalization of AGM postulates for revision. Section 6.2 presents the construction. Section 6.3 proves that the construction is fully characterized by certain set of postulates in distributive, compact logics. Section 6.4 concludes the chapter.

M. M. Ribeiro, *Belief Revision in Non-Classical Logics*,
SpringerBriefs in Computer Science,
DOI: 10.1007/978-1-4471-4186-0_6, © The Author(s) 2013

6.1 Generalized Postulates

By the same arguments presented in previous chapter, it is convenient to generalize revision postulates for the case when the input is a set of sentences. This generalization is straightforward and it was also presented in [Flo06]. Again the input A is considered to be a finitely representable set.

(**closure**) $K * A = Cn(K * A)$.
(**success**) $A \subseteq K * A$.
(**inclusion**) $K * A \subseteq K + A$.
(**vacuity**) If $K + A$ is consistent then $K * A = K + A$
(**consistency**) If A is consistent then $K * A$ is consistent.
(**extensionality**) If $Cn(A) = Cn(B)$ then $K * A = K * B$.

Consider also the following postulates borrowed from belief base literature, which will be useful in this chapter:

(**relevance**) If $\beta \in K \setminus K * A$ then there is K' such that $K \cap (K * A) \subseteq K' \subseteq K$ and $K' \cup A$ is consistent, but $K' \cup A \cup \{\beta\}$ is inconsistent.
(**uniformity**) If for all $K' \subseteq K$, $K' \cup A$ is inconsistent iff $K' \cup B$ is inconsistent then $K \cap K * A = K \cap K * B$

Relevance is a minimality postulate. It guarantees that in a revision only sentences that are somehow involved in the derivation of inconsistencies can be removed. *Uniformity* guarantees that two inputs in a revision over K that are inconsistent with the same subsets of K should give the same result.

Throughout this chapter a set A is considered inconsistent if it is trivial i.e. if $Cn(A) = \mathscr{L}$.

6.2 Construction

The strategy used to construct Revision Without Negation (RWN) is similar to the one used by Levi identity. First some sentences are removed from K so that when the input A is added the resulting belief set is consistent. To achieve this goal we will define the maximally consistent subsets of K with respect to A.

Definition 6.1 (*maximally consistent subsets of K w.r.t A*) [Del08] The maximally consistent subsets of K w.r.t. A form the set $K \downarrow A$ which is formally defined as follows. $X \in K \downarrow A$ iff:

1. $X \subseteq K$.
2. $X \cup A$ is consistent.
3. If $X \subset X' \subseteq K$ then $X' \cup A$ is inconsistent.

Selection function is defined as usual:

Definition 6.2 (*selection function*). [AGM85] A selection function for $K \downarrow A$ is a function γ such that:

- If $K \downarrow A \neq \emptyset$, then $\emptyset \neq \gamma(K \downarrow A) \subseteq K \downarrow A$.
- Otherwise, $\gamma(K \downarrow A) = \{K\}$.

Finally, *belief set revision without negation* (RwN) is defined as the intersection of those sets chosen by the selection function followed by an expansion:

$$K *_\gamma A = \left(\bigcap \gamma(K \downarrow A) \right) + A$$

The first result of this chapter shows that this construction satisfies AGM postulates for revision in any Tarskian and compact logic.

Lemma 6.3 [Del08] $K \downarrow A = K \downarrow B$ *iff for all* $X \subseteq K$, $X \cup A$ *is inconsistent iff* $X \cup B$ *is inconsistent.*

Proof Let $X \subseteq K$ and suppose that $X \cup A$ is inconsistent and that $X \cup B$ is consistent. By Lemma 6.6 there is X' such that $X \subseteq X' \in K \downarrow B$.

Since $X \cup A$ is inconsistent then, by monotonicity, $X' \cup A$ is also inconsistent and it follows that $X' \notin K \downarrow A$ which contradicts the hypothesis. □

Theorem 6.4 *Let the underlying logic be Tarskian and compact.* $K *_\gamma A = (\bigcap \gamma(K \downarrow A)) + A$ *satisfies the six basic AGM postulates for revision.*

Proof closure and success: Follows directly from the construction.

extensionality: Follows from Lemma 6.3.
vacuity: If $K + A$ is consistent then $K \downarrow A = \{K\}$, hence, $K *_\gamma A = K + A$.
consistency: Assume the consistency of A and the inconsistency of $\bigcap \gamma(K \downarrow A) + A$. Since $\bigcap \gamma(K \downarrow A) \subseteq X \in K \downarrow A$, by *monotonicity* $X \cup A$ is inconsistent, contradicting the definition. □

This construction also satisfies both relevance and uniformity if the underlying logic is Tarskian and compact.

Proposition 6.5 *If the underlying logic is compact then* $K *_\gamma A$ *satisfies relevance and uniformity.*

Proof relevance: Let $\beta \in K \setminus K *_\gamma A$, by compactness and Lemma 6.6 there is $X \in \gamma(K \downarrow A)$ such that $\beta \notin X + A$. It also holds that for any $X' \in \gamma(K \downarrow A)$, it holds that $\bigcap \gamma(K \downarrow A) + A \subseteq X' + A$. It follows that $K \cap (\bigcap \gamma(K \downarrow A) + A) \subseteq K \cap (X' + A)$. This holds in particular for the above mentioned X. Now just take K' as $K \cap (X + A)$ and verify that $K \cap (K * A) \subseteq K' \subseteq K$, $K' \cup A$ is consistent and $K' \cup A \cup \{\beta\}$ is not.

uniformity: Notice that for every $K' \subseteq K$, it holds that $K' \cup A$ is inconsistent iff $K' \cup B$ is inconsistent, then by Lemma 6.3, $K \downarrow A = K \downarrow B$, hence, $\bigcap \gamma(K \downarrow A) = \bigcap \gamma(K \downarrow B)$. □

6.3 Representation Theorems

In previous section it was proved that the construction of RwN satisfies AGM postulates. In this section we have two goals. The first goal is to show a set of postulates that characterizes RwN in any distributive and compact logic. The second goal is to show that AGM postulates characterizes RwN if the underlying logic satisfies AGM assumptions.

Showing a characterization that holds in any distributive and compact logic is not fully satisfactory, since many logics from Chap. 3 are not distributive. However, it is enough to characterize belief set revision in some interesting logics such as intuitionistic logic and certain Description Logics.

In distributive and compact logics vacuity and extensionality are redundant in the presence of relevance and uniformity.

Lemma 6.6 *In any compact logic if $X \subseteq K$ and $X \cup A$ is consistent, then there is a X' such that $X \subseteq X' \in K \downarrow A$.*

Proof This proof will be omitted because it is analogous with the proofs of Lemmas 5.9 and 4.3.   □

Proposition 6.7 *In the presence of* success *and* inclusion *it holds that:*

1. *Extensionality follows from uniformity.*
2. *Vacuity follows from relevance if the underlying logic is* distributive.

Proof

1. If $K + A$ is consistent then there is no $K' \subseteq K$ such that $K' + A$ is inconsistent. It follows, by *relevance* that there is no $\beta \in K \setminus K * A$. Hence, $K \subseteq K * A$. By *success* we have $K + A \subseteq K * A$ and by *inclusion* $K * A \subseteq K + A$.
2. If $Cn(A) = Cn(B)$ for all $K' \subseteq K$ then $K' + A$ is inconsistent iff $K' + B$ is inconsistent. It follows, by *uniformity*, that $K \cap K * A = K \cap K * B$. Since $Cn(A) = Cn(B)$, it follows that $(K \cap K * A) + A = (K \cap K * B) + B$. By *distributivity* $K + A \cap K * A + A = K + B \cap K * B + B$. By *success* and *inclusion* we have $K * A = K * B$.   □

Closure, success, inclusion, consistency, relevance and uniformity are enough to characterize RwN in any distributive and compact.

Representation Theorem 6.8 (RwN). *Let $\langle \mathscr{L}, Cn \rangle$ be compact and distributive logic. An operation $*$ over a belief set K is a revision without negation (RwN) iff is satisfies closure, success, inclusion, consistency, relevance and uniformity.*

Proof (**construction $\Rightarrow$ postulates**)

Theorem 6.4 showed that AGM postulates for revision are satisfied by the construction and proposition 6.5 proved that relevance and uniformity are satisfied.

   (**postulates $\Rightarrow$ construction**)

Let $*$ be an operator that satisfies AGM postulates and let γ be:

$$\gamma(K \downarrow A) = \{X \in K \downarrow A : K \cap (K * A) \subseteq X\} \quad \text{if } A \text{ is consistent}$$
$$= \{K\} \text{ otherwise.}$$

We will show that: (1) γ is well defined, (2) γ is a selection function and (3) $K *_\gamma A = K * A$.

1. By Lemma 6.3 and *uniformity* we have that γ is well defined, i.e., if $K \downarrow B = K \downarrow B$ then $\gamma(K \downarrow A) = \gamma(K \downarrow B)$.
2. If $K \downarrow A \neq \emptyset$ then A is consistent. It follows from *consistency* that $K * A$ is consistent. In this case, $(K * A) + A$ is consistent by *closure* and *success*. Since $(K \cap K * A) \cup A$, it follows from Lemma 6.6 that there is X such that $K \cap (K * A) \subseteq X \in K \downarrow A$. Hence $X \in \gamma(K \downarrow A)$.
3. If A is inconsistent (i.e. $Cn(A) = \mathscr{L}$) then, by *closure* and *success*, $K * A = K *_\gamma A = \mathscr{L}$.
   If A is consistent, $K \cap (K * A) \subseteq X$ for all $X \in \gamma(K \downarrow A)$. It follows that $K \cap (K * A) \subseteq \bigcap \gamma(K \downarrow A)$. By *monotonicity*, $(K \cap (K * A)) + A \subseteq \bigcap \gamma(K \downarrow A) + A$. It follows, by *distributivity*, that $K + A \cap K * A + A \subseteq K *_\gamma A$. By *success* and *inclusion* $K * A \subseteq K *_\gamma A$.
   Let $\beta \in \bigcap \gamma(K \downarrow A) \setminus K * A$. Since $\bigcap \gamma(K \downarrow A) \subseteq K$, by *relevance* there is K' such that $K \cap (K * A) \subseteq K' \subseteq K$, $K' \cup A$ is consistent and $K' \cup A \cup \{\beta\}$ is inconsistent. Since $K' \subseteq K$ and $K' \cup A$ is consistent, by Lemma 6.6 there is X such that $\beta \notin X$, $K' \subseteq X \in K \downarrow A$. Furthermore, $\beta \notin X$, because $K' \subseteq X$, $K' \cup A \cup \{\beta\}$ is inconsistent and $X \cup A$ is consistent. It follows that $\beta \notin X \in \gamma(K \downarrow A)$ and, hence, $\beta \notin \bigcap \gamma(K \downarrow A)$. We conclude that $\bigcap \gamma(K \bot A) \subseteq K * A$. It follows that $\bigcap \gamma(K \bot A) + A \subseteq K * A + A$ and, by *success* $K *_\gamma A \subseteq K * A$. □

The above result is useful for logics that are distributive, but not complemented. In Chap. 3 some examples of such logic were presented: intuitionistic logic and certain DLs such as $\mathcal{ALC}$ without ABoxes and $\mathcal{ALCO}$. Unfortunately, this result is not applicable to Horn logic and most DLs, specially those used in Semantic Web.

This section started announcing two goals. The first, already achieved, was to present a set of postulates that characterize RwN in distributive logics. The second was to prove that RwN is characterized by AGM postulates in logics satisfying the AGM assumptions. We will prove this second result showing that AGM postulates are equivalent to the postulates used to characterize RwN if the underlying logic satisfies AGM assumptions. For this purpose we will need a lemma proved by Makinson in [Mak87] and generalized here:

Lemma 6.9 *Let the underlying logic be Boolean and let A' be a complement of A. If $K + A = \mathscr{L}$ then $A' \subseteq Cn(K)$.*

Proof Since $K + A = \mathscr{L}$, we have that $A' \subseteq K + A$. It follows that $A' \subseteq K + A' \cap K + A$ and, by distributivity, $A' \subseteq Cn(K)$. □

Lemma 6.10 [Mak87] *If the underlying logic satisfies AGM assumptions then if the operation $*$ over a belief set K satisfies the AGM postulates for revision and A' is a complement of A then $*$ also satisfies the following:*
(Harper) There is an withdrawal $-$ such that $K - A' = (K * A) \cap K$.

Proof We will prove that $-$ satisfies every AGM postulate.

success: If $A' \not\subseteq Cn(\emptyset)$ then $Cn(A) \neq \mathscr{L}$ i.e. A is consistent. It follows by consistency of revision that $K * A$ is consistent.
By success of revision we have that $A \subseteq K * A$. It follows that $A' \not\subseteq K * A$ and, hence, $A' \not\subseteq K - A'$.
closure: Follows from closure of revision and the fact that the intersection of closed sets is closed.
inclusion: Follows directly from definition.
vacuity: If $A' \not\subseteq K$ then, by lemma 6.9, $K + A$ is consistent. By vacuity of revision we have that $K * A = K + A$. It follows that $K - A' = K$.
extensionality: If $Cn(A') = Cn(B')$ and B' is a complement of B then $Cn(A) = Cn(B)$ by Proposition 2.8. In this case $K * A = K * B$ by extensionality of revision. Hence $K - A' = K - B'$.
recovery: $(K - A') + A' = ((K * A) \cap K) + A'$. By distributivity we have $((K * A) + A') \cap (K + A')$. By success of revision $A \subseteq K * A$, hence, $(K * A) + A' = \mathscr{L}$. It follows that $K - A' = \mathscr{L} \cap (K + A') = K + A'$. We conclude that $K \subseteq K + A' = (K - A') + A'$. $\square$

Representation Theorem 6.11 *Let $\langle \mathscr{L}, Cn \rangle$ satisfy the AGM assumptions. An operator $*$ over a belief set K satisfies AGM postulates iff it satisfies: closure, success, inclusion, consistency, relevance and uniformity.*

Proof First we will prove that uniformity follows from vacuity, extensionality and closure and then we will prove that relevance is equivalent to Harper postulate. Since Harper postulate is redundant (Lemma 6.10) the proposition is proved.

1. Assume that for every set $K' \subseteq K$ we have that $K' + A$ is inconsistent iff $K' + B$ is inconsistent. We have two cases.
   If $K + A$ is consistent then $K * A = K + A$ and $K * B = K + B$ by *vacuity*. In this case $K = K \cap (K * A) = K \cap (K * B)$.
   If $K + A$ is inconsistent then, by closure and Lemma 6.9, $A' \subseteq K$ for a complement A' of A. By hypothesis $A' \cup B$ is inconsistent. By Lemma 6.9 again we have that $B' \subseteq Cn(A')$ and $A' \subseteq Cn(B')$ for a complement B' of B. It follows that $Cn(A') = Cn(B')$ and, by Proposition 2.8, $Cn(A) = Cn(B)$. From extensionality it follows that $K * A = K * B$ and, hence, $K \cap (K * A) = K \cap (K * B)$.

2. **(Harper $\Rightarrow$ relevance)**
   Since the underlying logic satisfies the AGM assumptions, by Proposition 2.14 and Theorem 5.16, we have that relevance for contraction follows from AGM postulates. By Harper postulate we have that there is an AGM contraction $K - A' = K \cap (K * A)$. It follows that if $\beta \in K \setminus K * A$ then there is K' such that $A' \not\subseteq Cn(K')$, but $A' \subseteq Cn(K' \cup \{\beta\})$. It follows that $Cn(K' \cup A \cup \{\beta\})$

is inconsistent and by the contrapositive of Lemma 6.9 $Cn(K' \cup A)$ is consistent.
(**relevance $\Rightarrow$ Harper**)

Since the underlying logic satisfies AGM assumption then, by Lemma 5.16, relevance for contraction and recovery are equivalent in the presence of the other AGM postulates for contraction. We will show that $K - A' = K \cap (K * A)$ satisfies closure, success, inclusion, extensionality, vacuity and relevance.

Closure and *inclusion* follows directly from construction. Since the negation is unique modulo equivalences (Proposition 2.7), by extensionality of revision, *extensionality* holds. By consistency and success of revision and Lemma 6.9 $A' \not\subseteq K * A$, hence, *success* holds. If $A' \not\subseteq K$, by Lemma 6.9, $K + A$ is consistent. It follows by vacuity of revision that $(K * A) \cap K = (K + A) \cap K = K$. Hence, *vacuity* of contraction also holds. Finally, to prove *relevance* notice that if $\beta \in K \setminus K - A'$ then $\beta \in K \setminus K * A$ and by relevance of revision there is K' such that $K - A' = K \cap (K * A) \subseteq K' \subseteq K$, $K' \cup A$ is consistent, but $K' \cup A \cup \{\beta\}$ is not. By Lemma 6.9 $A' \not\subseteq Cn(K')$ and $A' \subseteq Cn(K' \cup \{\beta\})$. $\square$

6.4 Conclusion

The revision in belief set presented in Chap. 4 assume the logic being closed under negation. However, several logics, such as Horn logic, don't satisfy this property. In this chapter it was presented a construction for belief set revision that doesn't depend on negation.

This construction was inspired by [Del08] and consists in first removing certain sentences from the belief set that together with the input would lead to a contradiction. After removing these sentences the input can securely be added.

This construction satisfies the AGM postulates for revision in any compact logic. Furthermore, if the underlying logic is well behaved then the AGM postulates fully characterizes this construction. This characterization is not easily generalized to a wider class of logics. A representation theorem was proved showing that a certain set of postulates characterizes the construction for logics that are compact and distributive. Unfortunately these properties are not satisfied by many logics listed in Chap. 3. Intuitionistic logic and certain specific DLs such as $\mathcal{ALCO}$ are exceptions i.e. they are examples of distributive and compact logics.

Most results in this Chapter are joint work with Wassermann and were first published in [RW09, RW10].

References

[AGM85]   Carlos Alchourrón, Peter Gärdenfors, and David Makinson. On the logic of theory change. *Journal of Symbolic Logic*, 50(2):510–530, 1985.

[Del08]   James P. Delgrande. Horn clause belief change: Contraction functions. In Gerhard Brewka and Jérôme Lang, editors, *Proceedings of the 10th International Conference* on the *Principles of Knowledge Representation and Reasoning (KR'08)*, pages 156–165, Sydney, Australia, Setembro 16-19 2008. AAAI Press.

[Flo06]   Giorgos Flouris. *On Belief Change and Ontology Evolution*. PhD thesis, University of Crete, 2006.

[Mak87]   David Makinson. On the status of the postulate of recovery in the logic of theory change. *Journal of Philosophical Logic*, 16:383–394, 1987.

[RW09]   Márcio Moretto Ribeiro and Renata Wassermann. AGM revision in description logics. In *Proceedings of the 1st Workshop on Automated Reasoning about Context and Ontology Evolution (ARCOE 2009)*, Pasadena, California, USA, Julho, 11-12 2009.

[RW10]   Márcio Moretto Ribeiro and Renata Wassermann. More about AGM revision in description logics. In *Proceedings of the 2nd Workshop Automated Reasoning about Context and Ontology Evolution (ARCOE 10)*, Lisboa, Portugal, Agosto, 16-17 2010.

Chapter 7
Base Revision in Logics Without Negation

Abstract Since many non-classical logics are not necessarily closed under negation, we need constructions for belief base revision that do not depend on this assumption. In this chapter, we present a list of six constructions for revision that do not depend on negation. Each construction is characterized by a set of postulates. Furthermore, the representation theorems that prove these characterizations hold in any compact logic.

Keywords Belief base theory · Partial meet contraction · Kernel contraction · Representation theorems · Negation · Non-classical logics

In Chap. 4, several constructions for belief base revision were presented. All these constructions assume that the underlying logic is monotonic, compact, and closed under negation that satisfies α-local non-contravention. In this chapter, constructions for belief base revision in logics that are not necessarily closed under negation are presented.

Classical belief base revision avoids contravention i.e., the presence of α and $\neg\alpha$ in the belief base. In semi-revision, we considered that there is a sentence $\perp$ that should be avoided. Now, we take a step further of generalization. It will be assumed the existence of a previously defined set Ω of "undesirable" sentences. Given the set Ω, a set A is Ω-consistent iff $Cn(A) \cap \Omega = \emptyset$. This generalization is powerful enough to take into account, for example, both notions of consistency in Description Logics. Furthermore, the revision operations in this chapter take finitely representable sets as inputs.

Six constructions for belief base revision are proposed, each one presented in a section. Figure 7.1 shows the organization of the sections in this chapter. Sections 7.1.1 and 7.1.2 present external kernel revision with strong success (EKRS) and with strong consistency (EKRC), respectively. Sections 7.1.3 and 7.1.4 present external partial meet revisions with strong success (EPMRS) and with strong consistency (EPMC), respectively. Section 7.2.2 presents internal kernel revision (IKR) and Sect. 7.2.1 presents internal partial meet revision (IPMR). The last section concludes the chapter.

M. M. Ribeiro, *Belief Revision in Non-Classical Logics*,
SpringerBriefs in Computer Science,
DOI: 10.1007/978-1-4471-4186-0_7, © The Author(s) 2013

	KernelRevision	PartialMeetRevision	
External	7.1.1	7.1.3	strong success
	7.1.2	7.1.4	strong consistency
Internal	7.2.2	7.2.1	

Fig. 7.1 Sections

7.1 External Revision Without Negation

External revision [Han93] consists of an expansion followed by a contraction:

$$(B + A) - \neg A$$

This construction, however, depends on the existence of the negation of A. It was already argued in Chap. 6, negation of a set of sentence is hard to define in certain logics, specially certain Description Logics.

To avoid the need of negation, constructions for revision that do not depend on it are presented. Our first approach is inspired in the idea behind semi-revision [Han97] (see Sect. 4.3.3).

Semi-revision construction consists in first expanding B by the input A and then removing the inconsistencies. In semi-revision, however, during the second step, the consolidation phase, elements of A may be removed. For this reason semi-revision do not satisfy success. To guaranty that success is satisfied, our strategy is to *protect* A in the consolidation phase.

If A is itself Ω-inconsistent, it is impossible to guaranty success and the consistency altogether. In this case, we have two options: prioritize success or prioritize consistency. If we opt to prioritize success then we can have only a weak version of consistency. On the other hand, if we opt to prioritize consistency we can guaranty only a weak version of success.

(**strong success**) $A \subseteq B * A$.
(**weak success**) If A is Ω-consistent then $A \subseteq B * A$.
(**strong consistency**) $B * A$ is Ω-consistent.
(**weak consistency**) If A is Ω-consistent then $B * A$ is consistent.

Section 4.3.3 showed two constructions for consolidation: kernel and partial meet. Since each of them can be combined with strong success or with strong consistency, we define four types of external revision:

(EKRS) external kernel revision with strong success.
(EKRC) external kernel revision with strong consistency.

(EPMRS) external partial meet revision with strong success.
(EPMRC) external partial meet revision with strong consistency.

Kernel revision is associated with core-retainment, while partial meet revision is associated with relevance.

(**core-retainment**) If $\beta \in B$ and $\beta \notin B * A$ there is a B' such that $B' \subseteq B \cup A$, B' is Ω-consistent, but $B' \cup \{\beta\}$ is not.

(**relevance**) If $\beta \in B$ and $\beta \notin B * A$ there is a B' such that $B * A \subseteq B' \subseteq B \cup A$, B' is Ω-consistent, but $B' \cup \{\beta\}$ is not.

The rest of the postulates must be generalized, but the generalization is straightforward:

(**inclusion**) $B * A \subseteq B + A$

(**pre-expansion**) $(B + A) * A = B * A$

Weak uniformity is not satisfied by the constructions in this chapter. This happens because in order to protect the input we violate the syntactic independence. In the following sections we define each of the four constructions and we prove a representation theorem for each of them.

7.1.1 External Kernel Revision With Strong Success

Kernel external revision is very similar to kernel semi-revision. It consists in adding the input and then removing the inconsistencies using the kernel consolidation. The difference here lies on the incision function which protects the input.

Before presenting the incision function, though, let us present a generalized version of kernel.

Definition 7.1 (*package kernel*) Let B be a belief base in $\mathscr{L}$ and let Ω the set of undesirable sentences of $\mathscr{L}$. The kernel $B \perp\!\!\!\perp_p \Omega$ is the set whose elements X are minimal subsets of B that imply some element of Ω i.e., $X \in K \perp\!\!\!\perp_p \Omega$ iff:

1. $\Omega \subseteq B$ (B is a subset of K).
2. $\Omega \cap Cn(B) \neq \emptyset$ (that imply some element of Ω).
3. if $B' \subset \Omega$ then $\Omega \cap Cn(B') = \emptyset$ (and is minimal).

Moreover, we need the following lemma:

Lemma 7.2 *Let B be a set of sentences in a compact logic $\langle \mathscr{L}, Cn \rangle$ and let B' be a subset of B and β be an element of B such that B' is Ω-consistent, but $B' \cup \{\beta\}$ is not. There is a set $X \subseteq B' \cup \{\beta\}$ such that $\beta \in X \in B \perp\!\!\!\perp_p \Omega$*

Proof This proof is omitted because it is very similar to the proof of Lemma 4.21. □

Incision function σ is a function that chooses at least on element of each α-kernel to be removed from the base. An *incision function that protects any input* chooses at least one element of each α-kernel, but never chooses elements of A. If A is itself

Ω-inconsistent then there will be an α-kernel contained in A. In this case, and only in this case, there will be an α-kernel with no elements chosen by σ.

Definition 7.3 (*incision function that protects any input*) A incision function σ that protects any input is a function σ over B that for every finitely representable A we have:

1. $\sigma(A, (B + A) \perp\!\!\!\perp_p \Omega) \subseteq \bigcup((B + A) \perp\!\!\!\perp_p \Omega)$.
2. If A is Ω-consistent and $\emptyset \neq X \in (B + A) \perp\!\!\!\perp_p \Omega$ then $X \cap \sigma(A, (B + A))$
3. $A \cap \sigma(A, (B + A) \perp\!\!\!\perp_p \Omega) = \emptyset$

A *external kernel revision with strong success* (EKRS) is defined using an incision function that protects the input σ as:

$$B * A = (B + A) \setminus \sigma(A, (B + A) \perp\!\!\!\perp_p \Omega)$$

This construction is characterized by inclusion, pre-expansion, core-retainment, strong success, and weak consistency. Core-retainment is associated with the fact that we used kernel contraction to remove Ω-inconsistencies. Strong success and weak consistency are related to the use of an incision function that protects any input.

Representation Theorem 7.4   (EKRS) *Let $\langle \mathcal{L}, Cn \rangle$ be Tarskian and compact. An operation $*$ over a belief base B is a external kernel revision with strong success (EKRS) iff is satisfies strong success, weak consistency, inclusion, pre-expansion, and core-retainment.*

Proof (**construction** $\Rightarrow$ **postulates**) Let σ be an incision function that protects any input and let $B *_\sigma A = (B + A) \setminus \sigma(A, (B + A) \perp\!\!\!\perp_p \Omega)$.

inclusion and pre-expansion: Follows directly from construction.
strong success and weak consistency: Follows from the definition of σ.
core-retainment: Let $\beta \in B \setminus B *_\sigma A$, by construction we have that $\beta \in \sigma(A, (B + A) \perp\!\!\!\perp_p \Omega)$. It follows that there is $X \in (B + A) \perp\!\!\!\perp_p \Omega$ with $\beta \in X$. Let $B' = X \setminus \{\beta\}$, we have that $B' \subseteq B \cup A$, B' is Ω-consistent and $B' \cup \{\beta\}$ is not.

(**postulates** $\Rightarrow$ **construction**) Let $*$ be an operation over B that satisfies the postulates mentioned and let σ be a function such that for every finitely representable A we have:

$$\sigma(A, (B + A) \perp\!\!\!\perp_p \Omega) = B + A \setminus (B + A * A)$$

We will show that (1) σ is an incision function that protects the input and (2) $B * A = B *_\sigma A$.

1. We have to show that the three conditions in Definition 7.3 are satisfied. Let $\beta \in \sigma(A, (B + A) \perp\!\!\!\perp \Omega)$, we have then that $\beta \in B \setminus (B + A) * A$ and by *pre-expansion* $\beta \notin B * A$. By *core-retainment* there is $B' \subseteq B + A$ such that B' is Ω-consistent, but $B' \cup \{\beta\}$ is not. It follows, by Lemma 7.2, that there is some $B'' \subseteq B'$ such that $B'' \cup \{\beta\} \in (B + A) \perp\!\!\!\perp_p \Omega$ and, hence, $\beta \in \bigcup(B + A \perp\!\!\!\perp_p \Omega)$.

Let A be a Ω-consistent set and let X such that $\emptyset \neq X \in (B + A) \perp \!\!\! \perp_p \Omega$. Assume by contradiction that $X \cap \sigma(A, (B + A) \perp \!\!\! \perp_p \Omega) = \emptyset$, then $X \subseteq B * A$. Since X is Ω-inconsistent, by *monotonicity* $B * A$ is Ω-inconsistent which contradicts *weak consistency*.

By *strong success* we have that $A \subseteq B * A$. It follows that $A \cap \sigma(A, (B + A) \perp \!\!\! \perp_p \Omega) = \emptyset$.

2. By definition and *pre-expasion*:

$$\sigma(A, (B + A) \perp \!\!\! \perp_p \Omega) = (B + A) \setminus ((B + A) * A)$$
$$= (B + A) \setminus (B * A)$$

It follows that:

$$B *_\sigma A = (B + A) \setminus \sigma(A, (B + A) \perp \!\!\! \perp_p \Omega)$$
$$= (B + A) \setminus ((B + A) \setminus (B * A))$$
$$= B * A \quad \text{by inclusion} \qquad \qquad \qquad \square$$

7.1.2 External Kernel Revision with Strong Consistency

Previous construction prioritizes success rather than consistency. The following construction prioritizes consistency.

In this construction it will be used an incision function that protects consistent inputs. An incision function that protects consistent inputs chooses at least one element of each α-kernel. If the input A is itself inconsistent there will be an α-kernel contained in A. In this case, A would not be protected.

Definition 7.5 (*incision function that protects consistent inputs*) A incision function σ that protects consistent input is a function σ over B that for every finitely representable A we have:

1. $\sigma(A, (B + A) \perp \!\!\! \perp_p \Omega) \subseteq \bigcup ((B + A) \perp \!\!\! \perp_p \Omega)$.
2. If $\emptyset \neq X \in (B + A) \perp \!\!\! \perp_p \Omega$ then $X \cap \sigma(A, (B + A))$
3. If A is Ω-consistent then $A \cap \sigma(A, (B + A) \perp \!\!\! \perp_p \Omega) = \emptyset$

The external kernel revision with strong consistency (EKRC) is defined exactly as the previous construction, but with an incision function that protects consistent inputs:

$$B * A = (B + A) \setminus \sigma(A, (B + A) \perp \!\!\! \perp_p \Omega)$$

This construction is characterized by inclusion, pre-expasion, core-retainment, weak success, and strong consistency. Strong consistency is guaranteed by the incision function that protects consistent inputs, since this function always chooses at least one element of each α-kernel.

Representation Theorem 7.6 (EKRC) *Let $\langle \mathcal{L}, Cn \rangle$ be Tarskian and compact. An operation $*$ over a belief base B is a external kernel revision with strong consistency (EKRC) iff is satisfies weak success, strong consistency, inclusion, pre-expansion and core-retainment.*

Proof (**construction $\Rightarrow$ postulates**) Let σ be a function that protects consistent input over a belief base B and let $B *_\sigma A = (B + A) \setminus \sigma(A, (B + A) \perp\!\!\!\perp_p \Omega)$.

inclusion and pre-expansion: Follow directly from construction.
weak success and strong consistency: Follow from the definition of the incision function that protects consistent inputs.
core-retainment: Let $\beta \in B \setminus B *_\sigma A$. By construction $\beta \in \sigma(A, (B + A) \perp\!\!\!\perp_p \Omega)$. It follows that $\beta \in X \in (B + A) \perp\!\!\!\perp_p \Omega$. Let $B' = X \setminus \{\beta\}$, then we have that $B' \subseteq B \cup A$. Since X is minimally Ω-consistent, B' is Ω-consistent, but $B' \cup \{\beta\}$ is not.

(**postulates $\Rightarrow$ construction**) Let $*$ be an operation that satisfies the postulates mentioned above and let σ be a function over B such that for every finitely representable A satisfies:

$$\sigma(A, (B + A) \perp\!\!\!\perp_p \Omega) = (B + A) \setminus ((B + A) * A)$$

We will prove that (1) σ is a incision function that protects consistent inputs and (2) $B * A = B *_\sigma A$.

1. We have to show that the three conditions in Definition 7.5 are satisfied. The first condition follows exactly as in Representation Theorem 7.4.

   Let $\emptyset \neq X \in (B + A) \perp\!\!\!\perp_p \Omega$ and assume by contradiction that $X \cap \sigma(A, (B + A) \perp\!\!\!\perp_p \Omega) = \Omega$. It follows that $X \subseteq B * A$. Since X is Ω-inconsistent, $B * A$ is also
   Ω-inconsistent by *monotonicity*. However, this contradicts *strong consistency*.
   Now suppose that A is Ω-consistent. By *weak success* $A \subseteq B * A$, hence, $A \cap \sigma(A, (B + A) \perp\!\!\!\perp_p \Omega)$.

2. This part follows exactly like Representation Theorem 7.4.　　　　　$\square$

7.1.3 External Partial Meet Revision with Strong Success

In previous sections, we showed how to consolidate a belief base using kernel technique i.e., removing elements from minimal Ω-inconsistent subsets. In this and the following sections, we will present an orthogonal strategy that consists in choosing certain maximal Ω-consistent subsets.

Remember that we defined revision with respect to a set Ω of undesired sentences. A set B is Ω-consistent iff no element of Ω can be inferred from B i.e., $\Omega \cap Cn(B) = \emptyset$. Notice that this goal is different from the goal of success in previous contraction operations where removing one sentence from the input was enough. For this reason, we need a different definition of remainder set here:

Definition 7.7 (*package remainder set*) Let B be a belief base in $\mathscr{L}$ and let Ω the set of undesirable sentences of $\mathscr{L}$. The remainder set $B \perp_p \Omega$ is the set whose elements X are maximal subsets of B that does not imply any element of Ω i.e., $X \in B \perp_p \Omega$ iff:

1. $\Omega \subseteq B$ (B is a subset of K).
2. $\Omega \cap Cn(B) = \emptyset$ (that does not imply any element of Ω).
3. if $\Omega \subset B' \subseteq B$ then $\Omega \cap Cn(B') \neq \emptyset$ (and is maximal).

Moreover, we need an upper-bound lemma for package remainder set.

Lemma 7.8 [AM81] *If $\langle \mathscr{L}, Cn \rangle$ is compact then for every $B \subseteq \mathscr{L}$, every $X \subseteq B$ and every $\Omega \subseteq \mathscr{L}$ for which $\Omega \not\subseteq Cn(X)$, there is a X' such that $X \subseteq X'$ and $X' \in B \perp_p A$.*

Proof This proof follows the same pattern as the ones of Lemmas 5.9, 4.3 and 6.6 □

To guaranty strong success a special selection function will be used. The *selection function that protects any input* chooses only elements from the remainder set that contains A. If A is itself Ω-inconsistent this will be impossible. In this case, the selection function will return the original belief base. Formally:

Definition 7.9 (*selection function that protects any input*) A selection function that protects any input γ is function over B that for any finitely representable A :

1. If A is Ω-consistent then $\emptyset \neq \gamma(A, (B + A) \perp_p \Omega) \subseteq (B + A) \perp_p \Omega$ and $A \subseteq \bigcap \gamma(A, (B + A) \perp_p \Omega)$.
2. Otherwise $\gamma(A, (B + A) \perp_p \Omega) = \{B\}$.

An external partial meet revision with strong success is defined using a selection function that protects any input γ as:

$$\bigcap \gamma(A, (B + A) \perp_p \Omega)$$

This construction is fully characterized by inclusion, pre-expansion, relevance, strong success, and weak consistency. Relevance is associated with partial meet method to remove Ω-inconsistencies and strong success is associated with the restrictions imposed to the selection function.

Representation Theorem 7.10 (EPMRS) *Let $\langle \mathscr{L}, Cn \rangle$ be Tarskian and compact. An operation $*$ over a belief base B is a external partial meet revision with strong success (EPMRS) iff is satisfies strong success, weak consistency, inclusion, pre-expansion and relevance.*

Proof (**construction ⇒ postulates**)

inclusion, pre-expansion and weak consistency: Follow from construction.

success: If A is Ω-consistent then $B * A = \bigcap \gamma(A, (B+A)\bot_p\Omega)$. By compactness and Lemma 7.8 we have that there is $X \in (B+A)\bot_p\Omega$ such that $A \subseteq X$.

If A is Ω-inconsistent then $\gamma(A, (B+A)\bot_p\Omega) = \{B\}$ and, hence, $B * A = B + A$.

relevance: If A is Ω-consistent then $B * A = \bigcap \gamma(A, (B+A)\bot_p\Omega)$. Consider $\beta \notin \bigcap \gamma(A, (B+A)\bot_p\Omega)$, there is $B' \in \gamma(A, (B+A)\bot_p\Omega)$ with $\beta \notin B'$. By definition, B' is Ω-consistent, $B' \cup \{\beta\}$ is not and $\bigcap \gamma(A, (B+A)\bot_p\Omega) \subseteq B' \subseteq B + A$.

If A is Ω-inconsistent then $\gamma(A, (B+A)\bot_p\Omega) = \{B\}$ and relevance is vacuously satisfied.

(**postulates ⇒ construction**) Let $*$ be an operation that satisfies the postulates mentioned in this theorem and let:

$$\gamma(A, (B+A)\bot_p\Omega) = \{X \in (B+A)\bot_p\Omega : B * A \subseteq X\} \text{ if } A \text{ is } \Omega - \text{consistent.}$$
$$= \{B + A\} \text{ otherwise.}$$

We will prove (1) that γ is a selection function that protects any input and (2) $B *_\gamma A = B * A$.

1. If A is Ω-consistent then $\gamma(A, (B+A)\bot_p\Omega) \neq \emptyset$. By *weak consistency, inclusion* and Lemma 7.8 there is $B * A \subseteq (B+A)\bot_p\Omega$
2. Notice that $\emptyset \neq B * A \subseteq X$ for every $X \in (B+A)\bot_p\Omega$ and that $B * A \subseteq \bigcap \gamma(A, (B+A)\bot_p\Omega)$.

If A is Ω-inconsistent then by *strong success, inclusion* and *relevance* we have that $B * A = B + A$.

If A is Ω-consistent then $\bigcap \gamma(A, (B+A)\bot_p\Omega) \subseteq B * A$ by *relevance* and $A \subseteq \bigcap \gamma(A, (B+A)\bot_p\Omega)$ by *strong success*. $\square$

7.1.4 External Partial Meet Revision with Strong Consistency

Selection function that protects any input do not guaranty that Ω-inconsistencies will be removed if the input A is itself Ω-inconsistent. To guaranty strong consistency we will define a *selection function that protects consistent inputs*. Differently from the previously defined selection function, this one always chooses at least one element of the remainder set. To guaranty weak success, if A is Ω-consistent then selection function that protects consistent inputs must select only sets that contains A.

Definition 7.11 (*selection function that protects any input*) A selection function that protects any input γ is function over B that for any finitely representable A:

1. $\emptyset \neq \gamma(A, (B+A)\bot_p\Omega) \subseteq (B+A)\bot_p\Omega$.
2. If A is Ω-consistent then $A \subseteq \bigcap \gamma(A, (B+A)\bot_p\Omega)$.

External partial meet revision with strong consistency is also defined as follows:

$$\bigcap \gamma(A, (B + A)\perp_p\Omega)$$

However, in this case the selection function γ protects only consistent inputs.

This construction is fully characterized by the same postulates that characterize previous construction with strong success and weak consistency exchanged by weak success and strong consistency.

Representation Theorem 7.12 (EPMRC) *Let* $\langle \mathscr{L}, Cn \rangle$ *be Tarskian and compact. An operation* $*$ *over a belief base B is a external partial meet revision with strong consistency (EPMRC) iff is satisfies weak success, strong consistency, inclusion, pre-expansion and relevance.*

Proof (**construction** $\Rightarrow$ **postulates**) This part follows exactly as in Representation Theorem 7.10.

(**postulates** $\Rightarrow$ **construction**) Let $*$ be an operation that satisfies the postulates mentioned in this theorem:

$$\gamma(A, (B + A)\perp_p\Omega) = \{X \in (B + A)\perp_p\Omega : B * A \subseteq X\}$$

We will prove (1) that γ is a selection function that protects consistent inputs and (2) $B *_\gamma A = B * A$

1. By consistency, inclusion and Lemma 7.8 we have that $B * A \subseteq B' \in (B + A)\perp_p\Omega$. It follows that $\gamma(A, (B + A)\perp_p\Omega) \neq \emptyset$. Furthermore, it follows from *weak success* that if A is Ω-consistent then $A \in \gamma(A, (B + A)\perp_p\Omega)$
2. Notice that $\emptyset \neq B * A \subseteq X$ for every $X \in (B + A)\perp_p\Omega$, hence, $B * A \subseteq \bigcap \gamma(A, (B + A)\perp_p \Omega)$. $\bigcap \gamma(A, (B + A)\perp_p \Omega) \subseteq B * A$ follows from *relevance*. Furthermore, if A is Ω-consistent then by *weak success* $A \in \bigcap \gamma(A, (B + A)\perp_p\Omega)$. $\qquad\square$

7.2 Internal Revision Without Negation

External revision consisted in first adding an new set of sentences A to the belief base B (expansion) and then removing the Ω-inconsistencies (consolidation). Internal revision uses a different strategy. First, some sentences are removed from the base B in order to "open space" to the input A which is added afterwards. Formally, internal revision is obtained given a contraction using Levi identity:

$$B * A = (B - \neg A) + A$$

The problem with internal revision is also the dependence on the existence of a negation for A. This section presents constructions for internal revision which does not depend on negation. The strategy used here is the same used in Chap. 6. First sentences from the base B that are Ω-inconsistent together with the input A are removed and then the result is expanded by A.

To guaranty minimality of change, the least amount of information should be lost when to "open space" for A. Again we have two options: use maximal subsets of B that are Ω-consistent with A (internal partial meet revision) or minimal Ω-inconsistent subsets (internal kernel revision).

Internal revision do not satisfy pre-expansion. Furthermore, both constructions that will be presented satisfy uniformity. The following is a generalized version of this postulate.

(**uniformity**) If for every $B' \subseteq B$ we have that $B' \cup A$ is Ω-inconsistent iff $B' \cup A'$ is Ω-inconsistent then $B \cap B * A = B \cap B * A'$.

The following sections presents two constructions for internal revision that do not depend on negation: Internal Partial Meet Revision without Negation (IPMRwN) and Internal Kernel Revision without Negation (IKRwN). Both constructions are followed by their respective representation theorem.

7.2.1 Internal Partial Meet Revision

The construction of internal partial meet revision is very similar to the construction of AGM revision without negation (see Chap. 6). The construction consists in the removal of certain sentences using the construction $B \downarrow A$ followed by the expansion by A:

$$B * A = \left(\bigcap \gamma(B \downarrow A) \right) + A$$

The only difference between this construction and the one presented in Chap. 6 is that now B is an arbitrary set of sentences i.e., a belief base. For this reason, closure is not satisfied. In fact, in terms of the postulates that characterize the construction, this is the only difference between them. The postulates that characterize this construction are: strong success, weak consistency, inclusion, relevance and uniformity.

The representation theorem proved in Chap. 6 holds if the underlying logic is compact and distributive. The following representation theorem, in contrast, holds for any compact logic i.e., it holds for a much wider class of logics which includes all logics listed in Chap. 3.

Representation Theorem 7.13   (IPMRwN) *Let $\langle \mathscr{L}, Cn \rangle$ be Tarskian and compact. An operation $*$ over a belief base B is a internal partial meet revision (IPMR) iff is satisfies strong success, weak consistency, inclusion, relevance, and uniformity.*

Proof (**construction $\Rightarrow$ postulates**)

strong success, weak consistency and inclusion: Follow directly from construction.

uniformity: Follows from Lemma 6.3.

relevance: Let $B *_\gamma A = \bigcap \gamma(B \downarrow A) + A$ and $\beta \in B \setminus B *_\gamma A$, then there is $X \in \gamma(B \downarrow A)$ with $\beta \notin X$. By definition of $B \downarrow A$ we have that $B \cap B *_\gamma A \subseteq X \subseteq B$, X is Ω-consistent, but $X \cup A \cup \{\beta\}$ is not.

(**postulates $\Rightarrow$ construction**) Let $*$ be an operation that satisfies the postulates mentioned in this theorem.

$$\gamma(B \downarrow A) = \{X \in B \downarrow A : B \cap (B * A) \subseteq X\} \text{ if } A \text{ is } \Omega\text{-consistent.}$$
$$= \{B\} \text{ otherwise.}$$

We will prove (1) that γ is a selection function and (2) that $B *_\gamma A = B * A$.

1. If $B \downarrow A = B \downarrow A'$ then by *uniformity* and Lemma 6.3 we have that $\gamma(B \downarrow A) = \gamma(B \downarrow A')$ i.e., γ is well defined.

   If A is Ω-consistent, by *weak consistency* $B * A$ is also Ω-consistent. By *inclusion* and *strong success* $(B \cap (B * A)) \cup A$ is Ω-consistent and by Lemma 6.6 there is $X \in B \downarrow A$ such that $B \cap (B * A) \subseteq X$. It follows that $\gamma(B \downarrow A) \neq \emptyset$.

2. $B \cap (B * A) \subseteq X$ for every $X \in \gamma(B \downarrow A)$. It follows that $B \cap (B * A) \subseteq \bigcap \gamma(B \downarrow A)$. Now notice that *inclusion* implies that $B * A \setminus A = B \cap (B * A)$. It follows that $(B * A) \setminus A \subseteq \bigcap \gamma(B \downarrow A)$, hence, $B * A \subseteq B *_\gamma A$.

   Let $\beta \in \bigcap \gamma(B \downarrow A) \setminus B * A$. Since $\bigcap \gamma(B \downarrow A) \subseteq B$, by *relevance* there is B' such that $B \cap (B * A) \subseteq B' \subseteq B$, $B' \cup A$ is Ω-consistent, but $B' \cup A \cup \{\beta\}$ is not. By Lemma 6.6 there is $X \in B \downarrow A$ such that $B' \subseteq X$ and $\beta \notin X$. It follows that $B \cap (B * A) \subseteq B \subseteq X \in B \downarrow A$ and, hence, $X \in \gamma(B \downarrow A)$. Since $\beta \notin X$, we have that $\beta \notin \bigcap \gamma(B \downarrow A)$. However, this contradicts the definition of β. We conclude that there is no $\beta \in \bigcap \gamma(B \downarrow A) \setminus B * A$ i.e., $B *_\gamma A \subseteq B * A$. $\square$

7.2.2 Internal Kernel Revision

Like internal partial meet revision, internal kernel revision consists in first removing certain sentences from the belief base B and then expanding by A. In internal kernel revision sentences from each of the minimal subsets of B that are Ω-inconsistent with the input A are removed during the first step.

Definition 7.14. The minimal subsets of B that are Ω-inconsistent with A forms the set $B \downdownarrows A$ defined as follows. $X \in B \downdownarrows A$ iff:

1. $X \subseteq B$. (X is a subset of B)
2. $X \cup A$ is Ω- inconsistent. (which is Ω-inconsistent with A)
3. If $X' \subset X$ then $X' \cup A$ is Ω-consistent. (and is minimal)

Incision function is defined as usual:

Definition 7.15 Let B be a belief base. An incision function for B is a function σ such that for every finitely representable set A :

1. $\sigma(B \downdownarrows A) \subseteq \bigcup B{\perp\!\!\!\perp} A.$
2. If $\emptyset \neq X \in B \downdownarrows A$ then $X \cap \sigma(B \downdownarrows A) \neq \emptyset.$

The difference between IPMRwN and IKRwN is analogous with the difference between kernel and partial meet contraction in Sect. 4.3. Not surprisingly the postulates that characterize IPMRwN are the same that characterize RIK with relevance exchanged by core-retainment.

Lemma 7.16 *Let B and A be sets of sentences in a compact logic $\langle \mathcal{L}, Cn \rangle$ and let B' be a subset of B and β be an element of B such that $B' \cup A$ is Ω-consistent, but $B' \cup A\{\beta\}$ is not. There is a set $X \subseteq B' \cup \{\beta\}$ such that $\beta \in X \in B \downdownarrows \Omega$*

Proof This proof is omitted because it is very similar to the proof of Lemma 4.21.
□

Lemma 7.17 *Let $\langle \mathcal{L}, Cn \rangle$ be a compact and monotonic logic. $B \downdownarrows A = B \downdownarrows A'$ iff for every $B' \subseteq B$ we have that $B \cup A$ is Ω-inconsistent iff $B \cup A'$ is also Ω-inconsistent.*

Proof ($\Rightarrow$) Suppose by contradiction that there is $B' \subseteq B$ such that $B' \cup A$ is Ω-inconsistent, but $B' \cup A'$ is not. By the previous lemma there is $X \subseteq B'$ such that $X \in B \downdownarrows A$. Since $B' \cup A'$ is Ω-consistent, $X \cup A'$ is also Ω-consistent by monotonicity. It follows that $X \notin B \downdownarrows A$ contradicting the hypothesis.

($\Leftarrow$) Suppose by contradiction that $X \in B \downdownarrows A$, but $X \notin B \downdownarrows A'$. We have two cases:

1. $X \cup A'$ is Ω-consistent. In this case, $X \cup A'$ is Ω-consistent, while $X \cup A$ is not which contradicts the hypothesis.
2. $X \cup A'$ is Ω-inconsistent. In this case, since $X \notin B \downdownarrows A'$ there must be $X' \subset X$ such that $X' \cup A'$ is Ω-inconsistent. It follows that $X' \subset X \in B \downdownarrows A$ and, hence, $X' \cup A$ is Ω-consistent contradicting the hypothesis.   □

Representation Theorem 7.18 (IKRwN) *Let $\langle \mathcal{L}, Cn \rangle$ be Tarskian and compact. An operation $*$ over a belief base B is a internal kernel revision (IKR) iff is satisfies strong success, weak consistency, inclusion, core-retainment and uniformity.*

Proof (**construction $\Rightarrow$ postulates**)

strong success, weak consistency and inclusion: Follow from construction.
core-retainment: Let $\beta \in B \setminus (B * A)$. It follows that $\beta \in \sigma(B \downdownarrows A)$, hence, there is X such that $\beta \in X \in \sigma(B \downdownarrows A)$. As consequence $(X \setminus \{\beta\}) \cup A$ is Ω-consistent, but $X \cup A$ is not.

(**postulates $\Rightarrow$ construction**) Let $*$ be an operation that satisfies the postulates mentioned in this theorem.

$$\sigma(B \downdownarrows A) = B \setminus (B * A)$$

Table 7.1 Belief base revision without negation

Postulates	EKRS	EKRC	EPMRS	EPMRC	IKRwN	IPMRwN
Success	Strong	Weak	Strong	Weak	Strong	Strong
Consistency	Weak	Strong	Weak	Strong	Weak	Weak
Inclusion	Yes	Yes	Yes	Yes	Yes	Yes
Pre-expansion	Yes	Yes	Yes	Yes	No	No
Minimality	Core-ret.	Core-ret.	Relev.	Relev.	Core-ret.	Relev.
Uniformity	No	No	No	No	Yes	Yes

Let $B *_\sigma A = (B \setminus \sigma(B \downarrow A)) \cup A$. We will prove (1) that σ is an incision function and (2) that $B *_\sigma A = B * A$.

1. If $B \downdownarrows A = B \downdownarrows A'$ then by *uniformity* and Lemma 7.17 $B * A = B *_\sigma A$ i.e., the function is well defined.

   If $\beta \in \sigma(B \downdownarrows A)$ then $\beta \in B \setminus (B * A)$. By *core-retainment* there is $B' \subseteq B$ such that $B' \cup A$ is Ω-consistent, but $B' \cup A \cup \{\beta\}$ is not. It follows, by Lemma 7.16, that there is $X \subseteq B$ such that $X \cup \{\beta\} \in B \downdownarrows A$ and, hence, $\beta \in \bigcup B \downdownarrows A$.

   Let $\emptyset \neq X \in B \downdownarrows A$ and assume by contradiction that $X \cap \sigma(B \downdownarrows A) = \emptyset$. It follows that $X \subseteq B * A$ and, since $X \cup A$ is Ω-inconsistent, it follows by monotonicity that $B * A$ is Ω-inconsistent. Since $\emptyset \neq X \in B \downdownarrows A$, we have that A is Ω-consistent. It follows by *weak consistency* that $B * A$ is Ω-consistent which is a contradiction. We conclude that $X \cap \sigma(B \downdownarrows A) \neq \emptyset$.

2.

$$B *_\sigma A = B \setminus \sigma(B \downdownarrows A) \cup A$$
$$= B \setminus (B \setminus (B * A)) \cup A$$
$$= (B \cap (B * A)) \cup A$$
$$\subseteq (B * A) \cup A$$
$$\subseteq B * A$$

The last line follows by *strong success*.

$$B *_\sigma A = (B \cap (B * A)) \cup A$$
$$= ((B * A) \setminus A)$$

The last line follows by *inclusion*. By *strong success*, we have that $B * A \subseteq B *_\sigma A$. $\qquad \square$

7.3 Conclusion

Constructions for base revision presented in Sect. 4.3 hold only if the underlying logic is closed under negation. In Chap. 3, we showed that several interesting logics are not closed under negation. One possible solution for this problem is semi-revision

[Han97], since semi-revision do not depend on negation. Semi-revision, however, fails to satisfy success which may be desirable. In this Chapter, it was presented how to revise a belief base in logics that are not closed under negation.

Six constructions for belief base revision that do not depend on negation were presented. Each construction is characterized by a set of postulates and, as proved by the representation theorems, this characterization holds for any Tarskian and compact logic. The main results of this Chapter are summed up in Table 7.1.

Most results in this chapter are joint work with Wassermann and some of the results were published first in [RW09].

References

[AM81]   Carlos Alchourrón and David Makinson. Hierarchies of regulation and their logic. In Hilpinen, editor, *New studies in deontic logic*, pages 125–148. D. Reidel Publishing Company, 1981.

[Han93]   Sven Ove Hansson. Reversing the Levi identity. *Journal of Philosophical Logic*, 22:637–669, 1993.

[Han97]   Sven Ove Hansson. Semi-revision (invited paper). *Journal of Applied Non-Classical Logics*, 7(2), 1997.

[RW09]   Márcio Moretto Ribeiro and Renata Wassermann. Base revision for ontology debugging. *Journal of Logic and Computation*, 19:721–743, 2009. Special Issue: Recent Advances in Ontology Dynamics.

Chapter 8
Algorithms for Belief Bases

Abstract Computationally the hardest part of revision and contraction operations in belief bases is finding the kernel/remainder set. In this chapter, we present algorithms for computing these sets. The similarities between the algorithms suggests that they are deeply related. We present this relation formally and show examples where computing the remainder set is much easier than computing the kernel and examples where the opposite is the case.

Keywords Remainder set · Kernel · Reiter's algorithm · Ontology debugging · Minimal cuts

In Chap. 7 it was shown how to adapt revision over belief base in order to apply it to logics that are not closed under negation. Computationally, the most difficult part of the construction is to compute the kernel or the remainder set. This chapter shows how to compute the kernel and the remainder set of a belief base w.r.t. an input sentence. These algorithms are adaptions of known algorithms from diagnosis [Rei87] and ontology debugging [SC03, Kal06].

Section 8.1 presents an algorithm and heuristics to find one element of the kernel. Section 8.2 defines minimal cuts and presents the Reiter algorithm. Sections 8.3 and 8.4 present algorithms for finding the kernel and the remainder set, respectively. These algorithms are clearly related with each other and Sect. 8.5 explores this relation. The last section concludes the chapter.

8.1 Computing One α-Kernel

The algorithm to find an α-kernel is pretty simple (see Algorithm 8.1). It consists in removing each element of B and testing if the resulting base still implies α. If this is not the case the element is reintroduced in B. The result of this process is a set $B' \subseteq B$ that do imply α which is minimal, i.e., an α-kernel.

M. M. Ribeiro, *Belief Revision in Non-Classical Logics*,
SpringerBriefs in Computer Science,
DOI: 10.1007/978-1-4471-4186-0_8, © The Author(s) 2013

Algorithm 8.1.1 Algorithm to find one α-kernel.

Kernel-black-box(B, α)
1. ▷ Find one element of the kernel $B \perp\!\!\!\perp \alpha$.
2. **for** $\beta \in B$
3.   **do if** $\alpha \in Cn(B \setminus \{\beta\})$
4.   **then** $B \leftarrow B \setminus \{\beta\}$
5. return B

Example 8.1 Assume that the underlying logic is CPL, let α in the input be q and B be the following set:

$$B = \{p, q, p \rightarrow q, r\}$$

1. The algorithm iterates over the elements of B (line 2). Let's assume that it iterates from left to right.
2. Checks that $q \in Cn(\{q, p \rightarrow q, r\})$ (line 3) so it removes p from B (line 3).
3. Checks that $q \notin Cn(p \rightarrow q, r)$ (line 3) so it does not change B.
4. Checks that $q \in Cn(q, r)$ (line 3), so $p \rightarrow q$ is removed (line 4).
5. Checks that $q \in Cn(q)$ (line 3), so it removes r (line 4).
6. Returns B which is $\{q\}$, an α-kernel of B (line 5).

Algorithm 8.1 is called black-box because it uses the theorem prover as a black box, i.e., it does not depend on it. A possible heuristic for this algorithm is to prune the base B before running it. This type of heuristics is based on the following result:

Proposition 8.2 *If* $B' \subseteq B$, $\alpha \in Cn(B')$ *and* $X \in B' \perp\!\!\!\perp \alpha$ *then* $X \in B \perp\!\!\!\perp \alpha$

Proof This proof is straightforward. Of course $\alpha \in Cn(X)$, because $X \in B' \perp\!\!\!\perp \alpha$. For the same reason there is no X' such that $X' \subset X$ and $\alpha \in Cn(X')$. Furthermore $X \subseteq B$, because $X \subseteq B' \subseteq B$. Hence $X \in B \perp\!\!\!\perp \alpha$.   □

From this result we conclude that if we run Algorithm 8.1 in any subset of the input B that implies α it will return an α-kernel of B. There are two well-known heuristics of this type.

The first is called expand-shrink (see Algorithm 8.1). It consists of an expansion phase of adding sentences of B in a set B' until B' implies the input α followed by a shrink phase (Algorithm 8.1).

Algorithm 8.1.2 Algorithm expand shrink.

Expand-shrink(B, α)
1. ▷ Heuristic to find one element of $B \perp\!\!\!\perp \alpha$
2. $B' \leftarrow \emptyset$
3. **for** $\beta \in B$
4.   **do** $B' \leftarrow B' \cup \{\beta\}$

5.   **if** $\alpha \in Cn(B')$
6.   ▷ Shrink
7.   **then** return Black-Box(B')

---

1. B' starts empty (line 2).
2. Iterates over the elements of B (line 3).
3. Inserts p into B' (line 4) and checks that $q \notin Cn(B')$ (line 5).
4. Inserts q into B' (line 4) and checks that $q \in Cn(B')$ (line 5).
5. Calls Algorithm 8.1 with inputs B' and α (line 7).

Example 8.3. Let the inputs B and α be the same of Example 8.1. The algorithm incrementally adds elements of B to an auxiliary set B' which is empty in the beginning and checks at each iteration if $\alpha \in Cn(B')$.

When Algorithm 8.1 is called $B' = \{p, q\}$, so $q \in Cn(B')$ and $B' \subseteq B$. By Proposition 8.2, Algorithm 8.1 with B' and p as inputs returns a p-kernel for B.

The second heuristic is called glass-box algorithm because it depends on the inference engine used. This algorithm consists of tracking down which sentences of B were used to infer α. These sentences forms a set $B' \subseteq B$ which implies α. After this step Algorithm 8.1 can be used. This heuristic, presented in [KPSH05], is very powerful because it prunes drastically the set B in just one run of the inference engine. However, the implementation of this algorithm is very difficult because it depends on the implementation of the inference engine.

8.2 Minimal Cuts

Algorithm 8.1 is used to find one α-kernel. In order to find the whole kernel we can use an idea presented by Wassermann in [Was00] which is based on Reiter algorithm [Rei87].

Reiter algorithm is used to find minimal cuts of a class of sets.[1]

Definition 8.4 (*Minimal Cut*) A cut for a class of sets $\mathcal{B}$ is a set B which intersects each set in the class $\mathcal{B}$. A cut C is minimal iff there is no cut C' properly contained in C i.e., if $C' \subset C$ then C' is not a cut for $\mathcal{B}$.

Reiter algorithm find all the minimal cuts of a class. Algorithm 8.2 is a simplified version of Reiter algorithm.

Algorithm 8.2.1 Reiter Algorithm to compute minimal cuts of a class

---

Reiter($\mathcal{B}$)
  1. $Cut \leftarrow \emptyset$
  2. $stack \leftarrow$ empty stack
  3. $S \leftarrow$ any element of $\mathcal{B}$

---

[1] We use here the term class as a synonym for set of sets.

4. **for** $s \in S$
5.   **do** insert $\{s\}$ in the beginning of $stack$.
6. **While** $stack$ not empty
7.   **do** $Hn \leftarrow$ last element of $stack$
8.   remove last element of $stack$
9.   **if** $\exists C \in Cut$ such that $C \subseteq Hn$
10.   **then** continue
11.   **elseIf** $\exists S \in \mathcal{B}$ such that $Hn \cap S = \emptyset$
12.   **then for** $s \in S$
13.   **do** insert $Hn \cup \{s\}$ in $stack$
14.   else
15. $Cut \leftarrow Cut \cup \{Hn\}$
16. return Cut

Variable Hn is a potential minimal cut of the kernel and it receives an element from the stack. If there is some cut which contains Hn then Hn is not minimal (lines 9 and 10) and Hn is discarded. Otherwise the algorithm checks if there is some element of $\mathcal{B}$ which does not intersect Hn. If there is none then Hn is a minimal cut and it is added to the set Cut of minimal cuts (line 15). Otherwise for each element s of the set that do not intersect Hn we have that $Hn \cup \{s\}$ is a potential minimal cut of $\mathcal{B}$. Every $Hn \cup \{s\}$ is added to the stack in order to be verified in another iteration of the loop. By the end of this process Cut contains every minimal cut of $\mathcal{B}$.

8.3 Computing the Kernel

To find the other elements of the kernel an adaptation of Reiter's algorithm is applied. Algorithm 8.3 is a simplified version of this algorithm.

Algorithm 8.3 begins extracting one α-kernel using the previously presented algorithms. At each iteration the least subset S of $B \setminus Hn$ that implies α is computed. If there is no cut which is a subset of Hn then Hn is a cut of the kernel. Otherwise S is an α-kernel.

Algorithm 8.3.1 Algorithm that computes the kernel $B \perp\!\!\!\perp \alpha$.

Kernel(B, α)
  1. $\triangleright$ Find the kernel $B \perp\!\!\!\perp \alpha$
  2. $Cut \leftarrow \emptyset$
  3. $stack \leftarrow$ empty stack
  4. $S \leftarrow Kernel - Black - Box(B, \alpha)$
  5. $Kernel \leftarrow \{S\}$
  6. **for** $s \in S$
  7.   **do** insert $\{s\}$ in the top of the $stack$.
  8. **while** $stack$ not empty
  9.   **do** $Hn \leftarrow$ last element of $stack$

10.		remove last element of $stack$
11.		**if** $\exists C \in Cut$ such that $C \subseteq Hn$
12.		**then** continue
13.		**elseIf** $\alpha \in Cn(B \setminus Hn)$
14.		**then** $S \leftarrow$ Black-Box$(B \setminus Hn, \alpha)$
15.		$Kernel \leftarrow Kernel \cup \{S\}$
16.		**for** $s \in S$
17.		**do** insert $Hn \cup \{s\}$ in the top of the $stack$
18.	**else**
19.		$Cut \leftarrow Cut \cup \{Hn\}$
20.	return $Kernel$

Example 8.5 Consider again the same inputs of Example 8.1.

1. The algorithm starts calculating one α-kernel of B using any of the algorithms presented in Sect. 8.3 and putting the result in S (lines 2–4). Following Example 8.1 we have that $S = \{q\}$.
2. The content of S is added to $Kernel$ (line 5).
3. Pushes $\{s\}$ into the $stack$ (line 6) for every element of $s \in S$. In our case the $stack$ contains $\{q\}$.
4. Hn gets the first element of the $stack$ ($\{q\}$), which is removed from the $stack$ (lines 9 and 10).
5. Checks if there is some cut smaller than Hn (line 11), which is not the case.
6. Checks if $\alpha \in Cn(B \setminus Hn)$ (line 13), which is the case.
7. Runs Algorithm 8.1 again and puts the result in S (line 14), which is now $\{p, p \rightarrow q\}$.
8. The content of S is again inserted in $Kernel$ (line 15).
9. For each $s \in S$ insert $Hn \cup \{s\}$ into the $stack$ (lines 16 and 17). Now the $stack$ contains $\{q, p\}$ and $\{q, p \rightarrow q\}$.
10. Repeats lines 9 to 11.
11. Now $Hn = \{q, p\}$ and, hence $\alpha \notin Cn(B \setminus Hn)$ (line 13).
12. Inserts $\{Hn\}$ into Cut (line 19).
13. Repeats steps 9 to 11.
14. $Hn = \{q, p\}$ and again $\alpha \notin Cn(B \setminus Hn)$ (line 13).
15. Inserts $\{Hn\}$ into Cut (line 19).
16. The $stack$ is now empty (line 8).
17. Returns $Kernel = \{\{q\}, \{p, p \rightarrow q\}\}$ (line 20).

8.4 Computing the Remainder Set

To find the remainder set we need to invert the algorithms presented above. Instead of removing elements from B and checking if α is still a consequence, the black-box algorithm for finding an element in the remainder set adds elements from B to a set B' and checks if α is still not a consequence.

Algorithm 8.4.1 Algorithm to find one element of $B \perp \alpha$ which includes X.

Remainder-black-box(B, α, X)
1. ▷ Find a set X' such that $X \subseteq X' \in B \perp A$.
2. **for** $\beta \in B$
3.   **do if** $\alpha \notin Cn(B' \cup \{\beta\})$
4.   **then** $B' \leftarrow B' \cup \{\beta\}$
5. return B'

To find the other elements of the remainder set an adaption of Reiter's algorithm is used (Algorithm 8.4).

Algorithm 8.4 to find the remainder set is very similar to Algorithm 8.3 to find the kernel. Hn is now a potential cut in the set $\{B \setminus Y : Y \in B \perp \alpha\}$. At each iteration if Hn is not yet a cut then the algorithm finds the elements of $B \perp \alpha$ which includes Hn.

8.5 Kernel and Remainder Sets

In previous section it was shown how to calculate the kernel $B \perp\!\!\!\perp \alpha$ (Algorithm 8.3) and the remainder set $B \perp \alpha$ (Algorithm 8.4). The similarity of these algorithms is evident. Intuitively, each of these algorithms is the converse of the other. There is in fact a close formal relation between them which was first studied in [FFKI06]. In this work, however, the authors focus in the relation between incision function and selection function. This section presents the formal relation between remainder set and kernel directly.

Algorithm 8.4.2 Algorithm that computes the remainder set $B \perp \alpha$.

Remainder(B, α)

1. ▷ Find the remainder set $B \perp \alpha$
2. $Cut \leftarrow \emptyset$
3. $stack \leftarrow$ empty stack
4. $S \leftarrow$ Remainder $-$ Black $-$ Box($B, \alpha, \emptyset$)
5. $Remainder \leftarrow \{S\}$
6. **for** $s \in B \setminus S$
7.   **do** insert $\{s\}$ in the top of the $stack$.
8. **while** $stack$ not empty
9.   **do** $Hn \leftarrow$ last element of $stack$
10.   remove last element of $stack$
11.   **if** $\exists C \in Cut$ such that $C \subseteq Hn$
12.   **then** continue
13.   **elseif** $\alpha \notin Cn(Hn)$

14.   **then** $S \leftarrow$ `Black − Box`(B, α, Hn)
15.   *Remainder* $\leftarrow$ *Remainder* $\cup \{S\}$
16.   **for** $s \in B \setminus S$
17.   **do** insert $Hn \cup \{s\}$ in the top of the *stack*
18.   **else**
19.   *Cut* $\leftarrow$ *Cut* $\cup \{Hn\}$
20. return *Remainder*

Lemma 8.6 *Let $\mathcal{B}$ be a class of sets and let $X \subseteq B$ for every $X \in \mathcal{B}$. Every minimal cut C of $\mathcal{B}$ satisfies $C \subseteq B$.*

Proof Suppose that C is a cut of $\mathcal{B}$ and that there is $\beta \in C$ such that $\beta \notin B$. Since $\beta \notin B$ and $X \subseteq B$ for each $X \in \mathcal{B}$ then β does not belong to any element of $\mathcal{B}$. It follows that $(C \setminus \{\beta\}) \cap X' \neq \emptyset$ for every $X' \in \mathcal{B}$. Since $S \setminus \{\beta\} \subset S$ then S cannot be a minimal cut which is an absurd. $\qquad\square$

Theorem 8.7 *Let $\langle \mathcal{L}, Cn \rangle$ be Tarakian and compact, $B \subseteq \mathcal{L}$ e $\alpha \in \mathcal{L}$:*

1. $X \in \{B \setminus Y : Y \in B \perp \alpha\}$ *iff* X *is a minimal cut for* $B \perp\!\!\!\perp \alpha$
2. $X \in B \perp\!\!\!\perp \alpha$ *iff* X *is a minimal cut for* $\{B \setminus X : X \in B \perp \alpha\}$

Proof

1. Since X is a minimal cut of $B \perp\!\!\!\perp \alpha$ by Lemma 8.6 we have that $X \subseteq B$. It follows that there is $Y \subseteq B$ such that $X = B \setminus Y$. We will show that $Y \in B \perp \alpha$.

- $Y \subseteq B$ trivially.
- Since $X \subseteq B$ and $Y = B \setminus X$, by simple set manipulation we have that $X = B \setminus Y$. Hence, by success of contraction we know that $\alpha \notin Cn(Y)$.
- If $X' \subset X$ then $\alpha \in Cn(B \setminus X')$ because X is a minimal cut of $B \perp\!\!\!\perp \alpha$ and, hence, there is $X'' \in B \perp\!\!\!\perp \alpha$ with $X' \cap X'' = \emptyset$. It follows that $X'' \subseteq B \setminus X'$ since $X'' \subseteq B$. We conclude that $\alpha \in Cn(Y \cup \{\beta\})$ for every $\beta \in B$ such that $\beta \notin X$.

2. Now let $Y \in B \perp \alpha$ and let $X = B \setminus Y$. By simple set manipulation we have that $Y = B \setminus X$. We will show that X is a minimal cut for $B \perp\!\!\!\perp \alpha$.

- Suppose that there is $X'' \in \perp\!\!\!\perp \alpha$ such that $X'' \cap X = \emptyset$. It follows that $X'' \subseteq B \setminus X$ and, hence, $\alpha \in Cn(B \setminus X) = Cn(Y)$ which violates the definition of Y. We conclude that for every $X'' \in B \perp\!\!\!\perp \alpha$ we have that $X'' \cap X \neq \emptyset$.
- Let $X' \subset X$, then $\alpha \in Cn(B \setminus X')$. By success of contraction, it follows that X' is not a cut for $B \perp\!\!\!\perp \alpha$. We conclude that X is a minimal cut for $B \perp\!\!\!\perp \alpha$. $\qquad\square$

This result shows that once the kernel (or the remainder set) is computed, the remainder (the kernel) can be extract without further calls to the theorem prover. Hence, we have two options to compute the remainder set, for example: use Algorithm 8.4 or first compute the kernel an then compute the remainder set using Theorem 8.7.

Although partial meet and kernel contractions are intimately related, Example 8.8 shows that in certain cases it is much faster to compute the kernel. Example 8.9, on the other hand, shows that in certain cases is much faster to compute the remainder set.

Example 8.8

$$
\begin{aligned}
B = \{ & b_1, b_1 \to a, \\
& b_2, b_2 \to a, \\
& \ldots, \\
& b_n, b_n \to a \}
\end{aligned}
$$

In this example $B \perp a$ has a number of elements proportional to 2^n, while $B \perp\!\!\!\perp a$ has a number of elements proportional to n.

Example 8.9

$$
\begin{aligned}
\mathrm{B} = \{ & b_1, b_1', \\
& (b_1 \vee b_1') \to b_2, (b_1 \vee b_1') \to b_2', \\
& (b_2 \vee b_2') \to b_3, (b_2 \vee b_2') \to b_3', \\
& \ldots, \\
& bn \vee b_{n'} \to a \}
\end{aligned}
$$

In this example $B \perp a$ has a number of element proportional to n, while $B \perp\!\!\!\perp a$ has a number of elements proportional to 2^n.

8.6 Conclusion

Computationally the hard part of the construction of base revision and contraction is finding the kernel and the remainder set. In this chapter, we showed algorithms for finding both the kernel and the remainder set of a belief base w.r.t. an input. These algorithms are adaptions of well-known algorithms from other artificial intelligence fields, namely, ontology debugging and diagnosis. More about these algorithms for belief debugging can be found in Kalyampur's thesis [Kal06] and Reiter algorithm can be found in [Rei87]. See [Was00] for more about how to apply these algorithms in belief base contraction.

The algorithms for finding the kernel and the remainder set are closely related. We showed that given one the other can be derived without further calls to the inference engine. The relation between kernel contraction and partial meet contraction was first studied in [FFKI06].

Besides this close relation, there are examples where the resulting kernel set is exponentially large while the resulting remainder set is only linearly large. On the other hand, there are examples where the remainder set is exponentially large and the kernel is linearly large. These examples show that for some instances it is faster to use kernel contraction while for others it is faster to use partial meet contraction. Whether it is possible to detect when it is better to use one or the other method is still an open problem.

References

[FFKI06] Marcelo A. Falappa, Eduardo L. Fermé, and Gabriele Kern-Isberner. On the logic of theory change: Relations between incision and selection functions. In Gerhard Brewka, Silvia Coradeschi, Anna Perini, and Paolo Traverso, editors, *Proceedings of the 17th European Conference on Artificial Intelligence (ECAI 2006)*, pages 402–406, Riva del Garda, Italy, 2006. IOS Press.

[Kal06] Aditya Kalyanpur. *Debugging and repair of owl ontologies*. PhD thesis, University of Maryland, College Park, MD, USA, 2006.

[KPSH05] Aditya Kalyanpur, Bijan Parsia, Evren Sirin, and James Hendler. Debugging unsatisfiable classes in OWL ontologies. *Journal of Web Semantics*, 3(4), 2005.

[Rei87] R. Reiter. A theory of diagnosis from first principles. *Artificial Intelligence*, 32(1): 57–95, 1987.

[SC03] Stefan Schlobach and Ronald Cornet. Non-standard reasoning services for the debugging of description logic terminologies. In Georg Gottlob and Toby Walsh, editors, *Proceedings of the Eighteenth International Joint Conference on Artificial Intelligence (IJCAI 03)*, pages 355–362, Acapulco, Mexico, 2003. Morgan Kaufmann.

[Was00] Renata Wassermann. An algorithm for belief revision. In Anthony G. Cohn, Fausto Giunchiglia, and Bart Selman, editors, *Proceedings of the 7th International Conference on Principles of Knowledge Representation and Reasoning (KR'00)*, pages 345–352, Breckenridge, Colorado, USA, Abril, 15-20 2000. Morgan Kaufmann.

Chapter 9
Conclusion

Abstract Most results in belief revision assumes that the underlying logic satisfies certain restrictive assumptions. We showed in this book several examples of logics that fail to satisfy these assumptions, e.g., most DLs, Horn logic, and intuitionistic logic. After that we presented ways of adapting classical belief revision in order for it to be compliant with a wider class of logics. In the case of belief set contraction we showed that this can be achieved by exchanging recovery by relevance in the AGM postulates. For belief base revision this can be achieved using selection/incision functions that protects the input. Finally, the most difficult case is belief set revision where the characterization is achieved only for distributive logics.

Keywords Belief revision · Non-classical logics · Ontology evolution · Representation theorems · Belief bases

Belief revision theory is a mature field of research and has applications in many fields such as philosophy, law, and specially computer science. Most results on belief revision, however, assume that the logic satisfies very restrictive properties, e.g., being supraclassical, closed under the standard connectives, etc.

Inspired by applications in semantic Web, in the past 10 years, some results showed that classical belief revision theory is not compliant with many logics [Flo06]. Several Description Logics, for example, was proved not to be compliant with AGM postulates for contraction [FPA05]. Some year after that Horn logics was also proved not compliant with AGM theory [LSST08]. As shown in the beginning of this book intuitionistic logic also adds to this list and probably many others.

However, many interesting applications in artificial intelligence would certainly profit with the use of belief revision techniques in Horn logic and Description Logics for example. Ontology evolution is a very good example. Many papers on ontology evolution seem to be ignoring years of development of belief revision theory. The equivalence between postulates and constructions proved by the Representation Theorem, for example, is a great abstraction that may give important insights into the field.

M. M. Ribeiro, *Belief Revision in Non-Classical Logics*,
SpringerBriefs in Computer Science,
DOI: 10.1007/978-1-4471-4186-0_9, © The Author(s) 2013

Table 9.1 Summary of representation theorem presented in the book

Operation	Representation theorem	Logical property
AGM contraction	4.9	AGM assumptions
AGM revision	4.12	AGM assumptions
Base PM contraction	4.19	Compactness
Base kernel contraction	4.24	Compactness
IKR	4.25	Compactness and α-local non contravention
EKR	4.26	Compactness and α-local non contravention
IPMR	4.27	Compactness and α-local non contravention
EPMR	4.28	Compactness and α-local non contravention
KSR	4.29	Compactness
PMSR	4.30	Compactness
Relevance-contraction	5.14	Compactness
RwN	6.8	Compactness and distributivity
EKRS	7.4	Compactness
EKRC	7.6	Compactness
EPMRS	7.10	Compactness
EPMRC	7.12	Compactness
IPMRwN	7.13	Compactness
IKRwN	7.18	Compactness

Although classical belief revision theory is not compliant with these and other logics, in this book it was shown that it is possible to adapt the postulates so that it became compliant. Our first step toward this adaption was to abstract the logic and deal only with the consequence operations which may satisfy certain properties.

In the case of contraction in belief sets this adaption can be achieved simply exchanging the postulate of recovery by relevance. We showed that this choice of postulates for belief set contraction has several advantages. Besides dealing more directly with minimality, relevance is compliant with any compact logic and it characterizes partial meet contraction in a wider class of logics.

Many results on belief base revision in non-classical logics were already presented in [HW02]. This work already showed representation theorems for base contraction that depend only in the logic being monotonic and compact. For belief base revision, however, the results depend on a strong assumption of the logic being closed under negation. Horn logic and most Description Logics are examples of logics that are not closed under negation. For this reason it was presented with several constructions for base revision that do not depend on negation. Each of these construction is fully characterized by a set of postulate and the representation theorems that prove these result assume only that the logic is Tarskian and compact.

Belief set revision is the most difficult scenario. It was presented a construction for it that does not depend on negation. This construction satisfies the AGM postulates for revision and this result assumes only that the underlying logic is Tarskian and compact. The representation theorem, however, makes a strong assumption on the logic being distributive.

Table 9.1 summarizes the results of Chaps. 5, 6 and 7. It shows for each representation theorem which logical properties was assumed.

Finally, it was also shown how to implement the operations of belief base contraction. The algorithms presented here are adaptions of algorithms known in other fields such as ontology debugging and diagnosis.

References

[Flo06]   Giorgos Flouris. *On Belief Change and Ontology Evolution.* PhD thesis, University of Crete, 2006.

[FPA05]   Giorgos Flouris, Dimitris Plexousakis, and Grigoris Antoniou. On applying the AGM theory to DLs and OWL. In Enrico Motta Yolanda Gil, V. Richard Benjamins, and Mark A. Musen, editors, *Proceedings of the 4th International Semantic Web Conference (ISWC 2005)*, pages 216–231, Galway, Ireland, November, 6-10 2005. Springer.

[HW02]   Sven Ove Hansson and Renata Wassermann. Local change. *Studia Logica*, 70(1): 49–76, 2002.

[LSST08]   Marina Langlois, Robert H. Sloan, Balázs Szörényi, and György Turán. Horn complements: Towards Horn-to-Horn belief revision. In Dieter Fox and Carla P. Gomes, editors, *Proceedings of the 23rd AAAI Conference on Artificial Intelligence (AAAI 2008)*, pages 466–471, Chicago, Illinois, USA, Julho 13-17 2008. AAAI Press.

Index